Informal Urban Street Markets

Through an international range of research, this volume examines how informal urban street markets facilitate the informal and formal economy not merely in terms of the traditional concerns of labor and consumption, but also in regards to cultural and spatial contingencies. In many places, street markets and their populace have been marginalized and devalued. At times, there are clear governance procedures that aim to prevent them, yet they continue to emerge even in the most institutionalized societies. This book gives serious consideration to what these markets reveal about urban life in a time of rapid, globalized urbanization and flows of people, knowledge and goods.

Clifton Evers is an Assistant Professor at the University of Nottingham, Ningbo, China.

Kirsten Seale is Adjunct Fellow at the Institute for Culture and Society at the University of Western Sydney.

Routledge Studies in Development and Society

For a full list of titles in this series, please visit www.routledge.com

Informal Urban Street Markets
International Perspectives

Edited by Clifton Evers and Kirsten Seale

Routledge
Taylor & Francis Group

LONDON AND NEW YORK

First published 2015 by Routledge

2 Park Square, Milton Park, Abingdon, Oxfordshire OX14 4RN
711 Third Avenue, New York, NY 10017

Routledge is an imprint of the Taylor & Francis Group,
an informa business

First issued in paperback 2018

Library of Congress Cataloging-in-Publication Data
Informal urban street markets : international perspectives / edited by
    Clifton Evers and Kirsten Seale.
        pages cm. — (Routledge studies in development and society ; 40)
    Includes bibliographical references and index.
    1. Street vendors—Developing countries.   2. Vending stands—
Developing countries.   3. Markets—Developing countries.   4. Informal sector (Economics)—Developing countries.   I. Evers, Clifton.
II. Seale, Kirsten.
    HF5459.D44I54 2015
    381'.18091724—dc23
    2014035555

ISBN: 978-1-138-79071-1 (hbk)
ISBN: 978-1-138-54639-4 (pbk)

Typeset in Sabon
by Apex CoVantage, LLC

Contents

PART II
Networks, Assemblages, and Territoriality

PART III
Service, Governance, and Policy

    Markets in Istanbul   149
    ASLI DURU

15 Shanghai's Unlicensed Taxis (*Hei Che*) as Informal
    Urban Street Market   158
    DUNFU ZHANG

16 Geographies of Unauthorized Street Trade and
    the "Fight Against Counterfeiting" in Milan   167
    KATE HEPWORTH

17 The Importance and Necessity of the Informal Market as
    Public Place in Delhi   176
    RANJANA MITAL

    Contributors   187
    Index   191

Figures

Tables

Acknowledgements

An edited collection such as this relies on a community of researchers. Our international research network looking at markets held workshops in Shanghai and Melbourne. We would like to thank participants from both workshops—Stephanie Hemelryk Donald, Ayxem Eli, Katie Hepworth, Lili Hernandez, Daphne Lowe Kelley, Maurizio Marinelli, Maša Mikola, Elspeth Probyn, Brian Risby, Sophie Watson, and Dunfu Zhang—for their contributions to a conversation about informal urban street markets. A special thank you to Stephi Donald for her extra support. These workshops were made possible through funding from the China in the World Centre at the Australian National University, RMIT University, and the University of Nottingham Ningbo China for which we are grateful. Thank you to Ann Deslandes, Joanne Watson, and the contributors to this collection for their collaboration. Thanks also to Routledge and editors Max Novick for seeing the potential in this project and for his advice and Jennifer E. Morrow for her editorial assistance and patience.

Nicholas and Louis Mariette, and Rosemary, Paul, and David Seale, have been fantastic cotravellers. Kat, Kajsa, and Lars Olausson have been wonderfully supportive, as have Westly and Lyn Evers.

Above all, we would like to acknowledge the communities who inspired this project and enable informal markets to happen across the globe.

1 Informal Urban Street Markets
International Perspectives

Kirsten Seale and Clifton Evers

Informal urban street markets are ubiquitous. Nevertheless, their existence can never be taken for granted and must be continually worked at. They are ambient constellations and complex changing configurations, always coming apart as they are coming together, always at the edge of themselves. Some endure. Others are provisional and ephemeral. In a city somewhere, someone knocks a hole in a brick wall one morning to provide access to private land cleared by a developer's machinery. For a few hours chickens are slaughtered, underwear sold, shoes repaired, and pensioners squat and slap their playing cards down on a plastic chair. By the evening the wall is bricked up, the police have moved everyone on, and the informal urban street market is gone forever.

This book brings together a collection of examples that address informal urban street markets across the globe. What we mean by 'example' follows on from the work of Giorgio Agamben (1993). For Agamben the example holds a unique status because it refers simultaneously to a vague and opaque generality, and a real case that is singular and refuses any claim to uniformity.

Informal urban street markets are made up of people buying and selling. Cash moves from pocket to hand to hand to pocket. Street food vendors sweat over open stoves. Illegal taxis compete for customers. Pharmaceuticals produced by multinationals are sold from collapsible tables. Stacked plastic products of all shapes and sizes lean and wobble on the back of carts. Socks and underwear are displayed on cloth and plastic sheets. Handbags dangle from a collapsible stand. A farmer slaughters a duck pulled from the cage mounted on the back of her pushbike. Middle-class shoppers turn away in horror and a tourist hurries to take a photo before the blood drains away. A restaurant owner buys vegetables in bulk from one vendor while the vendor's wife haggles with a retiree over the price of 200 grams of tofu. A migrant who has just arrived wanders through the stalls in the hope of meeting someone with the same cultural background.

Experiences of the informal economy in informal urban street markets go beyond labour, production, commodities, capital, and consumption. They facilitate the learning and passing on of information and skills (Geertz, 1978); generate "migrant cosmopolitanisms" (Kothari, 2008) and racism (Riccio,

1999); enable the mobility and participation of both privileged and marginalized social groups (Singerman, 1995; Usman, 2010; Williams, 2002); regenerate urban areas (Middleton, 2003); and in some circumstances they are central to the ongoing production of society and culture (Vecchio, 2013). Street markets are reliant on and produce social capital (Lyons & Snoxell, 2005), and vibrate with sociality (Watson, 2009).

INFORMAL MARKETS, INFORMAL ECONOMIES

Keith Hart's 1973 study on informal markets in urban Ghana defined the field. Since then scholarly research has canvassed an immense and diverse range of informal urban street markets in cities such as Hanoi (Lincoln, 2008), Yogyakarta (Gunadi, 2008), Mumbai (Anjaria 2006), Lusaka (Hansen, 2004), Nairobi (Lyons & Snoxell, 2005), Johannesburg (Cohen, 2010), Sana'a (Lauermann, 2013), Quito (Middleton, 2003), Baguio City (Milgram, 2011), Cusco (Bromley & Mackie, 2009), Mexico City (Crossa, 2009), Bogotá (Donovan, 2008), Barcelona (Kothari, 2008), London (Lyons, 2010), and Otara (de Bruin & Dupuis, 2000).

Understanding these sites is timely and necessary given that they are crucial in the emergence and reproduction of an informal economy that is, according to Robert Neuwirth, "in many countries—particularly in the developing world—[. . .]growing faster than any other part of the economy" (2011, p. 18). By 2020, the OECD projects that two-thirds of the workers of the world will be employed in the informal economy (ibid.). Neuwirth and others (Bhowmik, 2009; Coletto, 2010; Cook, 2008; Portes, Castells, & Benton, 1989) identify informal markets as key to economic growth and urban sustainability at the local, regional, national, and international scales. In many cities around the world informality—spatial, economic, social, political, bureaucratic—is the norm rather than the exception (Roy, 2004; 2009). Informality is fully integrated with cities. It is pertinent to all cities (AlSayyed, 2004).

Informality is not necessarily synonymous with poverty and the subaltern. The wealthy also engage in informal practices. There is 'informality from above'; for instance, to acquire and develop land through informal processes that devalorize "current uses and users and makes way for a gentrified future" (Roy, 2009, p. 84). There is also the example of consumption by the privileged through informal markets for reasons of "fun, sociality, distinction, discernment, the spectacular [. . .] rather than by economic necessity" (Williams, 2002, p. 1897).

There is no clear-cut binary of formal and informal. As Ann Varley (2013) explains, there is a "continuous urban fabric without obvious boundaries between 'formal' and 'informal'" (p. 13). In one way or another, they are linked (Chen, 2012; Daniels, 2004; Devas, 2001; Donovan, 2008;

Meagher, 2013; Robinson, 2006) or tangled (Meagher, 2010). Formal and informal urban practices are interwoven through deals between multiple informal and formal stakeholders and processes (Simone, 2001; Friedman, 2005). Rather than an informal–formal binary Varley (2013) points us towards the work of Ash Amin and Stephen Graham (1997), who suggest we understand cities as "as a set of spaces where diverse ranges of relational webs coalesce, interconnect and fragment" (p. 418). The relationships require ongoing negotiation, alliance building, restructuring, and resistance. Klarita Gërxhani (2004) explains that activity that falls within this sphere has diverse, complex, and ever-changing economic and noneconomic reasons and motives of existence, which include (but are not limited to) employment/unemployment, labor conditions in the formal sector, flexibility and satisfaction in work, culture and custom, skill sets, migration, allocation and rationing of goods, banning of goods and services, and the role—or lack thereof—of the State in regulating operations such as urban infrastructure, taxation, policy, health and safety rules, bureaucracy, and corruption. Participation may be necessary and by choice. At times activities in informal urban street markets are authorized and legitimate, but often they are not. Formal governance from state and city-sanctioned regulatory procedures folds and unfolds with unofficial and unregistered practices and encounters. This can mean that sometimes informal urban street markets are equated solely with criminality, illegality, chaos, spontaneity, and an absence of organization. However, informal urban street markets also involve degrees of planning, order, efficiency, deliberation, calculation, formalization, and regularization, even though at first glance these may appear lacking.

There can be a general tendency to ignore, repress, and regulate informal activities (Devas, 2001). When cities are 'cleaned up' as a demonstration of modernity and order, as well as the glorifying of the ability to do so (Kamete, 2007; 2009; Swanson, 2007), a clampdown on informal street markets and vendors usually takes place concurrently. Efforts to regulate, classify, contain, manage, and glorify urban order may perpetuate inequitable arrangements (Yiftachel, 2009a; 2009b). Inquiry into the regulation of informal street markets is often a concern of policy bodies such as local governments, business councils, and nongovernmental organizations. Martha Chen (2012) explains that they may turn a 'blind eye' to them or try to eliminate them, and that "most cities assign the 'handling' of street traders to those departments—such as the police—that deal with law and order" (p. 14). However, "Either stance has a punitive effect: eviction, harassment, and the demand for bribes by police, municipal officials and other vested interests" (ibid.). Some participants desire regularization and regulation, whilst others are wary (Chen, 2012; Kumar, 2012; Sinha & Roever, 2011). For example, traders' associations and organizations struggle to provide lasting impacts and a voice for vendors so they have legal and social protections in the face of complex local and global pressures (Brown, Lyons, & Dankoco, 2010; Dunn, 2014). They must work out what appropriate regulations should be as well as what should or should not be regulated: "A *missing* regulatory environment can be as costly to informal operators as an *excessive* regulatory environment" (Chen, 2012, p. 14). Consequently, informal urban street markets are frequently sites of conflict and compromise (Cross, 2000; O'Connor, 2000). They also involve resistance (Kamete, 2007; 2009; Kudva, 2009). Debates about the political, judicial, and economic rules of state-enforced planning and regulation can lead to empowerment and exploitation. According to Chen (2012), "There is a need to rethink regulations to determine what regulations are appropriate for which components of informal employment" (p. 14).

INFORMAL URBAN STREET MARKETS AS SUBALTERN URBANISM

Informal urban street markets can irritate government departments, urban planners, corporations, and real estate developers as they try to fix, determine, and claim territory over space and meaning in cities (Kudva, 2009). The political questions, problems, possibilities, and the inventions they disclose and animate challenge the hegemonic ideological work of subordination, centralization of power, homogenization, unification, normalization, and standardization.

Informal urban street markets can de-frame canonical knowledge of cities. Governmental blueprints, planning, strategies, and design are transformed, dissolved, and disorientated by the whirlwinds of idiosyncrasies, energetics, dislocations, hybridities, specificities, and partialities. There is sometimes the reconfiguring of trading networks, transport routes, labour, housing, electricity supply, sanitation, water supply, refuse collection, and other urban infrastructure. Top-down efforts to militate against and exclude nondominant modes of city living, city imaginaries, and becoming-city are resisted and alternatives are put forward and emerge. The public interest of the city is not left wholly in the hands of governmental bodies, the hegemonic political apparatus, private developers, and institutional financiers (Roy, 2009).

While there are clear governance procedures that aim to contain or prevent the operation of and participation in informal urban street markets, they continue to emerge even in the most institutionalized societies and economies because they can provide many benefits to city life: employment, skills training, transport, housing, health care, innovation, belonging, cultural exchange, well-being, community, urban regeneration, identity formation, place-making, recycling, and economic growth. New ways to live and exercise power are found. Gaps in formal provision of urban services can be filled (Meagher, 2007; Nunan & Satterthwaite, 2001). Informal does not simply equal powerlessness.

Mörtenböck and Mooshammer (2008) raise the issue of subject formation in informal street markets and suggest it is an open question for all participants as they invoke distinctive arrangements. People self-organize to enhance their labor situation, as well as gain access to resources and facilities (Motala, 2002; Simone, 2001). Unseen spaces become apparent (Chattopadhyay, 2012), and space is contested for with the state and private enterprise, and realized as sites of local transformations (Bromley & Mackie, 2009; Crossa, 2009; Donovan, 2008). Everyday technologies are reimagined. Bodies are rehabituated as newcomers lightly and intensely brush past or encounter those with a more established habitus. In other words, despite efforts to dominate, sideline, destroy, undermine, constrain, and coerce events so that they can be controlled, organized, and planned for, "new possibilities are ushered into being" through self-cultivation and self-education (Connolly, 2002, p. 1). There is the proliferating of possibilities through collective (human and more-than-human) action, adaptation, and experimentation (Gibson-Graham, 2006).

Informal urban street markets involve kinetic flows of bodies, emotions, goods, sounds, temperatures, smells, finance, psychologies, ideas, discourses, fauna, flora, waste, and, and, and . . . . Co-constitutive relationships (productive, restrictive, and potential) are formed with architecture, institutions, regulations, policy models, and city infrastructure. Informal urban street markets are more-than-human (Evers, this volume). They vibrate with materiality and literally matter (Bennett, 2010). They are always qualitatively different because at different times and under certain circumstances the sociological, biological, spatial, and psychological intermingle and interpenetrate. Geographer Doreen Massey calls this a "throwntogetherness" of negotiation, "judgement, learning, improvisation" (Massey, 2005, p. 162). Throwntogetherness isn't necessarily unplanned; rather, the different constituent parts work together to produce outcomes.

We adopt the concept of the rhizome from the work of Deleuze and Guattari (1987) to figure the informal urban street market. Unlike the arboreal model of the tree, the rhizome is an ever-changing multiple and decentralized process of human and nonhuman relationships that generates nodes, offshoots, clusters, and links that are always in a state of reconfiguration (at variable speeds) based on the given conditions of possibility. Alfredo Brillembourg and Hubert Klumpner (2005) in a study of Caracas have also figured the wider city as rhizome. They argue there is form, but this can be multiple and fluid, and tends to defy modern urban planning efforts. Kim Dovey (2012) similarly remarks upon how urban informality involves "complex adaptive assemblages."

Informal urban street markets have a tendency to morphology, being open ended, disruption, hybridity, adaptation, multiplication, and revision. As the chapters in this book elucidate, these sites carry the ability to produce new ontologies, representations, agencies, subjectivities, practices, places, and so forth. We recognize their potential for creativity. Informal urban street markets are *eventful* and orientated to being otherwise (Deleuze & Guattari, 1987). They have to be given the ever-shifting contingent relationships that need to be negotiated. Mind you, movement is not always free-flowing and is, at times, stymied and (re)directed. There are blockages and restrictions because of regulations, policies, and infrastructure. We should remember that equilibrium can and does take place, yet if we look close enough we can see even the most long-standing and established markets are sensitive to perturbations as new relationships come into play.

IDENTITY, BELONGING, AND SOCIALITY

Part I of the collection addresses the formation and constitution of identity, belonging, and sociality in the informal urban street market. Lelia Green interrogates the Vô Danh street market in Vietnam and learns from the women who work as street vendors about how such street markets are their only opportunity to work. Through sociality and cooperation the women find ways to negotiate various levels of authority whose aim remains to control the women's activity and order street life in ways that marginalize the women and diminish their ability to be part of the city. Despite such governance and precarious working conditions the street market and these women's ongoing efforts facilitate social and economic ties across local and global scales to engender an important web of interdependencies that supports life in the city.

Yusuf Abdulazeez and Sundramoorthy Pathmanathan's study of migrant workers in informal street markets in Sokoto, Nigeria, demonstrates how informal urban street markets facilitate a wide range of opportunities for social inclusion, as well as employment, that are otherwise unavailable to workers and residents living in urban environments. Their findings show that social contact and exchange in informal street market communities sustains extensive urban social ecologies in Sokoto, and beyond in other cities in Nigeria.

In Daisy Tam's piece informality itself produces a type of sociality. Tam's auto-ethnography of workers at London's Borough Market studies an informal economy that does more than merely replicate or shadow the official economy of the highly regulated market space and place which hosts it. This underground manifestation developed independent mobilities, rhythms, and encounters, many of them socially based. Similarly, in Tony Mitchell's account of belonging in the El Chopo street market in Mexico City, informal trade in subcultural and underground music enhances a sense of identity for its community, many of whom feel un(der)represented or marginalized in other urban spaces and places.

Tam's contribution to the book recognizes that the informal markets that nest within the formal infrastructure of urban markets produce alternative identities, belongings, and socialities to those promulgated in official narratives of the market. For Maša Mikola, informality in the market can be a mode beyond the economic or the spatial. In this case, it is also linguistic.

Mikola looks at how the minor language of minor literature, as understood by Gilles Deleuze and Félix Guattari, is deployed by market communities to produce an informality in language. The multiple knowledges and experiences of place enabled by this informality challenge the single grand narrative constructed in and for Melbourne's central Queen Victoria Market via dominant discourses of nationalism, parochialism, and multiculturalism. Through informal language, as well as informal practices and use of space, multiple and alternative articulations of identity, belonging, and sociality become possible.

Sophie Watson's (2006) research on London markets found that sociality and belonging in the market were susceptible to external social, political, and material factors. Micol Brazzabeni's ethnographic study on socially marginalized *cigano* sellers in Lisbon's open-air markets confirms that sociality and belonging in the market are responsive and reactive to wider social and political conditions. In Lisbon, Brazzabeni observed how *cigano* vendors situated themselves, and were situated, in relation to national identity in response to the 2007–08 Global Financial Crisis. This group of vendors deployed their identity as outsiders in Portuguese society, and with reference to wider national and global currents, to define a particular (advantageous) relationship to their customers, and to the products they were selling.

NETWORKS, ASSEMBLAGES, AND TERRITORIALITY

In Part II we map the overlapping networks of relationships that are informal urban street markets. Territories form and dissipate because of these networks. These territories are contexts, not simply places or spaces. They emerge and fade as the biological, sociological, and psychological assemble and disassemble. This process happens at varying speeds and intensities to yield outcomes. In Clifton Evers's analysis of the assemblage(s) of 'becoming-tricycle' and the Pengpu night market in Shanghai, these terms deliberately recall the work of Gilles Deleuze and Félix Guattari. Evers writes, "assemblages are an interplay of relationships *between* heterogeneous elements that bring out particular capacities of those elements and that yield 'articulations' [by which] I mean: properties, tendencies, momentums, forces, qualities, meanings, and behaviours." Evers places an emphasis on more-than-human agency when considering such assemblages. The hypotactic, open-ended 'and' of networks and assemblages is also emphasized in Emily Potter's work on formal and informal water markets in Chennai. Potter draws a diagrammatic example of an assemblage of the material and the biological and the temporal (clean drinking water and polluted water and plastic bottle and distribution channels and urban infrastructure and local, state and/or federal government, and . . . , and . . .) that is always coming together and coming apart.

The network of the informal urban street market becomes a territory through contingent relationships between assemblage and context. In the informal urban street market, space and place in the city is always in a process of becoming smooth (heterogenous, transgressive) and striated (ordered, regulated). In space striated by governance and policy, territoriality is the recourse of the informal. Informal markets are, literally and metaphorically, contested ground. In her discourse analysis of urban space and place at Rio de Janeiro's iconic Copacabana and Ipanema beaches, Kirsten Seale concludes that visible informality is in conflict with the place-image of Rio as discursively communicated in global and local media, political, and material economies. Yet in contrast to official crackdowns, criminalization, and moral panic about informal markets' use and occupation of urban space, residents and visitors are largely unconcerned at the presence of informal street markets on the beaches, and indeed actively support and sustain them through commercial and social transaction.

This section could equally be titled "Complexes, Constellations, Entanglements," in which there is always an explicit or implicit acknowledgement of a network, assemblage, or territory that deploys both informal and formal in the production of space, practice, knowledge, communication, and representation in the city. Networks, assemblages, and territoriality occur where formal and informal come together spatially, materially, socially, and perhaps become contiguous, transform the other or together, or grow rhizomatically into something new or distinct. Kiran Keswani and Suresh Bhagavatula's work acknowledges the nuances operating in the ever-shifting interrelation between the formal and informal in an authorized street market in Bangalore, where layers of informality are folded into the social and spatial economies of the market. Similarly Khalilah Zakariya's ethnographic research on the quotidian rhythms of a street vendor at Kuala Lumpur's night markets records the negotiations between formal and informal at the varying scales of the market stall, the street, and the city.

For Colin Marx (2009), the test for nuanced analysis and policy on informality is not to reinscribe the limitations of binarism and to situate the informal according to its own terms. Nashaat Hussein's piece on the participation of unlicensed street vendors in the collective event of the January 2011 Revolution in Cairo underscores the radical challenge to political, legal, and social discourse necessary for that shift to occur. Hussein links the social and political activism and objectives of the revolution to an overturning of an urban social and spatial order which stigmatizes and marginalizes vendors. The territoriality enacted by the protestors informally occupying the space and place of Tahrir Square achieves a recognition of informality that depends on the internal logic of informality for coherence.

SERVICE, GOVERNANCE, AND POLICY

Overall, service, governance, and policy in and for the 21st-century globalized city rarely heeds Marx's (2009) exhortation to conceptualize informality on its own terms. Where the formal is the dominant discourse, the informal is Other. This 'Other-ing' makes external governance of informal urban street markets problematic. In Part III, Ranjana Mital's work on the threats to informal street markets in the highly contested urban space of Delhi highlights the struggle for street traders' associations as the voice of the subaltern, to speak and be heard in the development of urban, legal, or economic policy that directly impacts on their livelihood.

Both Asli Duru and Katie Hepworth tie official policies on urban street markets to neoliberal political economy and modes of governance. In mapping the effects of the closure of markets in central Istanbul on women who are responsible for provisioning in their households or family groups, Duru ascribes the subsequent relocation of markets to peripheral urban areas to neoliberal commodification of public urban space. Hepworth connects a 'crackdown' on unauthorized street vendors selling counterfeit designer fashion in Milan's streets to economic protectionism by a network of formal economic interests at global and local scales that includes the owners of the brands being counterfeited, and local businesses, property owners, and residents.

Significantly, Hepworth documents how increased regulation of the unauthorized street traders led to a decreased number of visitors frequenting licensed businesses in shopping precincts where the street vendors had previously been prevalent, which in turn impacted financially on local businesses. The perception amongst consumers was that amenities and services in the areas were diminished when the informal street sellers were removed. This effect supports the view that informal markets rarely supplant the formal urban economy. Instead, they are adjunct, complementary, or supplementary, and augment what already exists by providing services and products that are otherwise unavailable or meeting a consumer need or demand that is neglected. Dunfu Zhang's study of the unlicensed taxi industry in Shanghai illustrates this. In spite of policies and governance which criminalize the drivers and provoke moral panic and anxiety about public safety, a market for the taxis persists because they provide transportation and employment where these options are limited.

WHAT DO INFORMAL MARKETS DO?

Following Ananya Roy (2009), we choose the title *informal* urban street markets because we are interested in informality as an epistemological concept that signifies generative activity through alternative knowledge and use of city services, land, housing, livelihood strategies, and so forth, thereby challenging top-down urban planning approaches. Informality here then is a "localised knowledge" as a "politics of redress" (Kudva, 2009, p. 1617), where global historical, social, economic, political, and ecological forces meet local instances of these forces during informal urban street markets.

By "informal" we refer to an adaptability that we believe to be crucial in conceptualizing and generating sustainable ways of living in cities, imagining cities, and becoming-cities. In this sense, informality lends itself to a counter-hegemonic politics. For us, it can signify 'misuse,' transgression, and reinterpretation in the face of hegemonically imposed planning prescriptions that build hierarchies of privilege in the city that favour private industry and state power (Roy, 2009). On the other hand, we should be cautious about "romantic depictions" of the informal economy (Williams & Round, 2008), "heroic" representations of informality (Varley, 2013), and aestheticizing poverty (Roy, 2005; Varley, 2013). Breman's (2013) study of the informal economy in India argues that the informal economy can equally involve oppressive labour conditions and exploitative debt arrangements. Others have noted the economic vulnerability of street vendors in the informal economy (Gunadi, 2008; de Bruin & Dupuis, 2000). Concerns about the effect of neoliberalism, international finance institutions, private developers, and economic globalization on the functioning of informal urban street markets are substantiated in research in this book and elsewhere (Carr & Chen, 2001; Dunn, 2014; Galemba, 2008; Owusu, 2007; Roy, 2009; Swanson, 2007; Walsh, 2010).

Informal urban street markets push back against such pressures, or manoeuvre to incorporate, redress, and rework these forces for their own advantage and to "make do" (Simone, 2001). Neoliberal macro-operations of power are undeniably designed to thwart the development of informal economies, while at the same time providing the very material, political, and social conditions necessary for the emergence of the entrepreneurialism and innovation of the informal street market (see Seale in this volume on the relationship between informal street markets and globalized neoliberal trade and labour). For the producers and consumers who rely on it, the informal urban street market shapes and serves multiple purposes at any given time.

Rather than ask or define what the informal urban street market means then, we are more interested in asking, what do informal urban street markets *do*? The research represented in this book is a turn to an ongoing thinking through and experiencing of what informal urban street markets do, or do not do, to come to terms with contemporary urbanization. The case studies in this book are embedded in and with the "dailyness" (Gibson-Graham, 2006) of informal urban street markets. This dailyness exists as a wellspring for the empirical investigations and questions in this book that re-vision what informal urban street markets do in the perpetuation and interruption of economies, subject positions, places, identifications, semiotics, bodies, sensations, routes, concepts, and more. The work here achieves a re-visioning through a wide range of disciplines, methodologies, readings, and contexts. What we learned during the process of editing this book is that informal urban street markets yield events co-constituted by historical, social, economic, political, and ecological forces—mundane *and* sensational, irregular *and* regular, subaltern *and* oppressive, commonplace *and* unfamiliar, banal *and* exciting, creative *and* unimaginative, spontaneous *and* calculated, old *and* new. Informal urban street markets can never be predicted or ascertained as a whole. They are a hurly-burly of interventions and proliferations.

REFERENCES

Agamben, G. (1993). *The coming community* (M. Hardt, Trans.). Minneapolis: University of Minnesota Press.

AlSayyad, N. (2004). Urban informality as a "new" way of life. In A. Roy & N. AlSayyad (Eds.), *Urban informality: Transnational perspectives from the Middle East, Latin America, and South Asia*. Lanham, MD: Lexington Books.

Amin, A., & Graham, S. (1997). The ordinary city. *Transactions of the Institute of British Geographers, 22*(4), 411–429.

Anjaria, J. S. (2006). Street hawkers and public space in Mumbai. *Economic and Political Weekly, 21*(41), 2140–2146.

Bennett, J. (2010). *Vibrant matter: A political ecology of things.* Durham, NC: Duke University Press.

Bhowmik, S. (2009). *Street vendors and the global urban economy.* New Delhi: Routledge.

Breman, J. (2013). *At work in the informal economy of India: A perspective from the bottom up.* London: Oxford University Press.

Brillembourg, A., & Klumpner, H. (2005). *Informal city: Caracas case.* Munich: Prestel.

Bromley, R.D.F., & Mackie, P. K. (2009). Displacement and the new spaces for informal trade in the Latin American city centre. *Urban Studies, 46*(7), 1485–1506.

Brown, A., Lyons, M., & Dankoco, I. (2010). Street traders and the emerging spaces for urban voice and citizenship in African cities. *Urban Studies, 47*(3), 666–683.

Carr, M., & Chen, M. A. (2001). *Globalisation and the informal economy: How global trade and investment impact on the working poor.* Background paper commissioned by the ILO Task Force on the Informal Economy. Geneva, Switzerland: International Labour Office. Available at http://www.ilo.org/employment/Whatwedo/Publications/WCMS_122053/lang--en/index.htm

Chattopadhyay, S. (2012). *Unlearning the city: Infrastructure in a new optical field.* Minneapolis: University of Minnesota Press.

Chen, M. A. (2012). *The informal economy: Definitions, theories and policies.* WIEGO Working Paper No. 1. Women in Informal Employment Globalizing and Organizing. Available at http://wiego.org/sites/wiego.org/files/publications/files/Chen_WIEGO_WP1.pdf

Cohen, J. (2010). How the global economic crisis reaches marginalised workers: The case of street traders in Johannesburg, South Africa. *Gender & Development, 18*(2), 277–289.

Coletto, D. (2010). Ambulantes and camelôs: The street vendors. In *The Informal Economy and Employment in Brazil: Latin America, Modernization, and Social Changes.* New York: Palgrave Macmillan.

Connolly, W. (2002). *Neuropolitics: Thinking, culture, speed.* Minneapolis: University of Minnesota Press.

Cook, D. (Ed.). (2008). *Lived experiences of public consumption: Encounters with value in marketplaces on five continents.* Basingstoke, UK: Palgrave Macmillan.

Cross, J. (2000). Street vendors, and postmodernity: Conflict and compromise in the global economy. *International Journal of Sociology and Social Policy, 20*(1), 29–51.

Crossa, V. (2009). Resisting the entrepreneurial city: Street vendors struggle in Mexico City's historic centre. *International Journal of Urban and Regional Research, 33*(1), 43–63.

Daniels, P. W. (2004). Urban challenges: The formal and informal economies in mega-cities. *Cities, 21*(6), 501–511.

De Bruin, A., & Dupuis, A. (2000). The dynamics of New Zealand's largest street market: The Otara Flea Market. *International Journal of Sociology and Social Policy, 20*(1–2), 52–73.

Deleuze, G., & Guattari, F. (1987). *A thousand plateaus: Capitalism and schizophrenia* (B. Massumi, Trans.). Minneapolis: University of Minnesota Press.

Devas, N. (2001). Does city governance matter for the urban poor? *International Planning Studies, 6*(4), 393–408.

Donovan, M. G. (2008). Informal cities and the contestation of public space: The case of Bogotá's street vendors, 1988–2003. *Urban Studies, 45*(1), 29–51.

Dovey, K. (2012). Informal urbanism and complex adaptive assemblage. *International Development Planning Review, 34*(4), 340–368.

Dunn, K. (2014). Street vendors in and against the global city: VAMOS Unidos. In R. Milkman & E. Ott (Eds.), *New Labor in New York: Precarious Worker Organizing and the Future of Unionism*. Ithaca: Cornell University Press.

Friedmann, J. (2005). Globalization and the emerging culture of planning. *Progress in Planning, 64*, 183–234.

Galemba, R. (2008). Informal and illicit entrepreneurs: Fighting for a place in the neoliberal economic order. *Anthropology of Work Review, 29*(2), 19–25.

Geertz, C. (1978). The bazaar economy: Information and search in peasant marketing. *American Economic Review, 68*(2), 28–32.

Gërxhani, K. (2004). The informal sector in developed and less developed countries: A literature survey. *Public Choice, 120*(3–4), 267–300.

Gibson-Graham, J. K. (2006). *A postcapitalist politics*. Minneapolis: University of Minnesota Press.

Gunadi, B. A. (2008). *Vulnerability of urban informal sector: Street vendors in Yogyakarta, Indonesia*. Munich Personal RePEc Archive.

Hansen, K. T. (2004). Who rules the streets? The politics of vending space in Lusaka. In K. T. Hansen & M. Vaa (Eds.), *Reconsidering informality: Perspectives from urban Africa*. Nordiska Afrikaninstitutet.

Hart, K. (1973). Informal income opportunities and urban employment in Ghana. *Journal of Modern African Studies, 11*(1), 61–89.

Kamete, A. Y. (2007). Cold-hearted, negligent and spineless? Planning, planners and the (r)ejection of "filth" in urban Zimbabwe. *International Planning Studies, 12*(2), 153–171.

Kamete, A. Y. (2009). In the service of tyranny: Debating the role of planning in Zimbabwe's urban "clean-up" operation. *Urban Studies, 46*(4), 897–922.

Kothari, U. (2008). Global peddlers and local networks: Migrant cosmopolitanisms. *Environment and Planning D: Society and Space, 26*(3), 500–516.

Kudva, N. (2009). The everyday and the episodic: the spatial and political impacts of urban informality. *Environment and Planning A, 41*(7), 1614–1628.

Kumar, R. (2012). *The regularization of street vending in Bhubaneshwar, India: A policy model*. WIEGO Policy Brief (Urban Policies) No. 7. Women in Informal Employment Globalizing and Organizing. Available from http://wiego.org/sites/wiego.org/files/publications/files/Kumar_WIEGO_PB7.pdf

Lauermann, J. (2013). Practicing space: Vending practices and street markets in Sana'a, Yemen. *Geoforum, 47*, 65–72.

Lincoln, M. (2008). Report from the field: Street vendors and the informal sector in Hanoi. *Dialectical Anthropology, 32*(3), 261–265.

Lyons, M. (2010). Temporary migration, the informal economy and structural change: London's bicycle rickshaw riders. *Local Economy, 22*(4), 376–387.

Lyons, M., & Snoxell, S. (2005). Creating urban social capital: Some evidence from informal traders in Nairobi. *Urban Studies, 42*(7), 1077–1097.

Marx, C. (2009). Conceptualising the potential of informal land markets to reduce urban poverty. *International Development Planning Review, 31*, 335–353.

Massey, D. (2005). *For space*. London: Sage.

Meagher, K. (2007). Introduction: Special issue on informal institutions and development in Africa. *Afrika Spectrum, 42*(3), 405–418.

Meagher, K. (2010). The tangled web of associational life: Urban governance and the politics of popular livelihoods in Nigeria. *Urban Forum, 21*(3), 299–313.

Meagher, K. (2013). *Unlocking the informal economy: A literature review on linkages between formal and informal economies in developing countries.* WIEGO Working Paper No. 27. Women in Informal Employment Globalizing and Organizing. Available at http://wiego.org/sites/wiego.org/files/publications/files/Meagher-Informal-Economy-Lit-Review-WIEGO-WP27.pdf

Middleton, A. (2003). Informal traders and planners in the regeneration of historic city centres: The case of Quito, Ecuador. *Progress in Planning, 59*, 71–123.

Milgram, B. L. (2011). Reconfiguring space, mobilizing livelihood: Street vending, legality, and work in the Philippines. *Journal of Developing Studies, 27*(3–4), 261–293.

Mörtenböck, P., & Mooshammer, H. (2008). Spaces of encounter: Informal markets in Europe. *Arq: Architectural Research Quarterly, 12*(3–4), 347–357.

Motala, S. (2002). *Organizing in the informal economy: A case study of street trading in South Africa.* SEED Working Paper No. 36. Geneva, Switzerland: International Labour Office. Available at www.ilo.org/wcmsp5/groups/public/@ed_emp/@emp_ent/@ifp_seed/documents/publication/wcms_117700.pdf

Neuwirth, R. (2011). *Stealth of nations: The global rise of the informal economy*. New York: Pantheon Books.

Nunan, F., & Satterthwaite, D. (2001). The influence of governance on the provision of urban environmental infrastructure and services for low-income groups. *International Planning Studies, 6*(4), 409–426.

O'Connor, A. (2000). *El Chopo rock market: The struggle for public space in Mexico City.* Available at http://ontario.indymedia.ca/twiki/bin/view/Toronto/ElChopoMarket

Owusu, F. (2007). Conceptualizing livelihood strategies in African: Planning and development implications of multiple livelihood strategies. *Journal of Planning Education and Research, 26*(4), 450–465.

Portes, A., Castells, M., & Benton, L. A. (1989). *The informal economy: Studies in advanced and less developed countries.* Baltimore, MD: Johns Hopkins University Press.

Riccio, B. (1999). Senegalese street-sellers, racism and the discourse on "irregular trade" in Rimini. *Modern Italy, 4*(2), 225–239.

Robinson, J. (2006). *Ordinary cities: Between modernity and development*. London: Routledge.

Roy, A. (2004). Transnational trespassings: The geopolitics of urban informality. In A. Roy & N. AlSayyad (Eds.), *Urban informality: Transnational perspectives from the Middle East, Latin America, and South Asia* (pp. 298–317). Lanham, MD: Lexington Books.

Roy, A. (2005). Urban informality: Towards an epistemology of planning. *Journal of the American Planning Association, 17*(2), 147–158.

Roy, A. (2009). Why India cannot plan its cities: Informality, insurgence and the idiom of urbanization. *Planning Theory, 8*(1), 76–87.

Simone, A. (2001). Straddling the divides: Remaking associational life in the informal African city. *International Journal of Urban and Regional Research, 25*(1), 102–117.

Singerman, D. (1995). *Avenues of participation: Family, politics and networks in urban quarters of Cairo.* Princeton, NJ: Princeton University Press.

Sinha, S., & Roever, S. (2011). *India's national policy on urban street vendors.* WIEGO Policy Brief (Urban Policies) No. 2. Women in Informal Employment Globalizing and Organizing. Available at http://wiego.org/sites/wiego.org/files/publications/files/Sinha_WIEGO_PB2.pdf

Swanson, K. (2007). Revanchist urbanism heads south: The regulation of Indigenous beggars and street vendors in Ecuador. *Antipode, 39*(4), 708–728.

Usman, L. M. (2010). Street hawking and socio-economic dynamics of nomadic girls of Northern Nigeria. *International Journal of Social Economics, 37*(9), 717–734.

Varley, A. (2013). Postcolonialising informality? *Environment and Planning D: Society and Space, 31*(1), 4–22.

Vecchio, G. (2013). *Tianguis* shaping *ciudad.* Informal street vending as a decisive element for economy, society and culture in Mexico. *Planum: The Journal of Urbanism, 1*(26).

Walsh, J. C. (2010). Street vendors and the dynamics of the informal economy: Evidence from Vung Tau, Vietnam. *Asian Social Science, 6*(11). Available at http://ccsenet.org/journal/index.php/ass/article/view/6672

Watson, S. (2006). *City publics: The (dis)enchantments of urban encounters.* London: Routledge.

Watson, S. (2009). The magic of the marketplace: Sociality in a neglected public space. *Urban Studies, 46*(8), 1577–1591.

Williams, C., & Round, J. (2008). A critical evaluation of romantic depictions of the informal economy. *Review of Social Economy, 66*(3), 297–323.

Williams, C. C. (2002). Why do people use alternative retail channels? Some case study evidence from two English cities. *Urban Studies, 39*(10), 1897–1910.

Yiftachel, O. (2009a). Critical theory and "gray space": Mobilization of the colonized. *City, 13*(2/3), 240–256.

Yiftachel, O. (2009b). Theoretical notes of "gray cities": The coming of urban apartheid? *Planning Theory, 8*(1), 88–100.

Part I

Identity, Belonging, and Sociality

2 "People Have to Find Their Own Way of Making a Living"

The Sale of Food in an Informal Ha Noi Street Market

Lelia Green

My first connection with street markets in Ha Noi assaulted my comfortable, Western middle-class values around food, hygiene, and the treatment of animals. As a personal experiment in health-promoting lifestyles, I haven't eaten meat, dairy, or eggs since 2007 (Green, Costello, & Dare, 2010). Seven years of plant-based eating have heightened my awareness of the human exploitation of other species and although such (mis)treatment occurs mainly behind closed doors in Western societies, in factory farms and abattoirs (Singer & Mason, 2006), it is visible in an agrarian economy like Vietnam; street market life offers a range of evidence.

This chapter was prompted by my initial reactions to slabs of raw meat presented for sale on splintered chopping boards; fish flapping and gasping on a shallow metal tray as their eyes clouded over; the plastic buckets full of soft mammalian brains and twisted offal; and the tubs of tiny crabs crawling over each other. As I and research assistant/interpreter Nguyen Hong Van (Van) started talking to the vendors, however, it was the vendors' lives which became fascinating to me: their relationship with authorities in the market, and with the social and financial imperatives of home.

This chapter is not a verified set of truths. One vendor told us that the market was contracting, another that it was expanding; some said it was busier than 10 years ago, others said the opposite. We didn't check these perceptions with police or the authorities, and we didn't challenge what people told us or attempt to verify it in any way other than through observation.

We were continually sensitive to the fact that the informal street market locale of our fieldwork operated in contravention of local laws and regulations. There was ample evidence of surveillance by authorities and some indication of corrupt regulatory practice, as is discussed shortly. A foreigner calling attention to the street market in any official way might have had uncomfortable consequences for the vendors who used it to sell their goods and services, and would certainly have made the market more identifiable. Challenging what we were told could have made interviewees fearful or suspicious, and I was guided in my responses by Van, who had helped other Western researchers in similar circumstances, although not in this market. Given that we did not challenge what we were told, there were, however, no obvious inconsistencies between what we saw in 50 hours of fieldwork, and what we were told, in terms of the quotes that follow.

It is the pavement-based street sellers, established and itinerant, whose presence makes the Vietnam streetscape particularly 'other' to the tourist gaze: "the all-seeing eye that demarcates the 'real, authentic locals' from everyone else" (Urry, 1992, p. 178). Whether squatting on the pavement for hours at a time, or walking constantly, yoked by a pole to two heavy baskets, street vendors are the central element of the informal Ha Noi street market, and particularly reveal it as different from equivalent contexts in the West. The physicality of the street vendors' work, their conditions, and the number of people engaged in selling all point towards a limited range of possible occupations and livelihoods, and the absence of a developed Western-style system of social security.

This dependence upon the informal economy is underscored by the challenges that street vendors face. Vietnamese authorities, for example, proclaimed a ban on all street vending in most of Ha Noi's Old Quarter from 1 July 2008. Police officers, local security forces, and ward officials are all responsible for moving street traders on, and keeping the streets and pavements clear. The local Ha Noi press will compare a street market with a messy village, arguing that the city itself is tidy and regulated, and no place for the untidy business of small scale street-based trade and barter. This is a classic demarcation between purity and danger (Douglas, 1966), where the pristine, middle-class city is polluted by the 'chaos' and poverty of the rural poor, implicit in much of the products and processes of street vendors' activities. Indeed, the implication is that authorities perceive itinerant vendors as "social pollution; [. . .] individuals or social groups in a particular location whose beliefs or actions are seen as 'polluting' " (Urry, 1992, p. 182). This is one reason why informal street markets in Ha Noi face closure. They reveal to visiting tourists that some Vietnamese people have few options and resources, challenging the pervasive message of prosperity, modernization, order, and growing wealth.

Although it attends to these wider issues, this chapter focuses on food selling; drawing upon interviews with 10 female food vendors from within the Vô Danh street market and two from outside. All names are aliases, including that of the Vô Danh street market itself. The chapter's reliance upon interviews with female participants reflects the focus upon food; Vietnamese men working in a street-based context prefer "motor cycle repair or sale of higher priced goods [. . . because] their earnings are higher" (Bhowmik, 2012, p. 38, citing Tantiwiramanond, 2004). As Nga, who sells sticky rice, says, "Only men who are too desperate sell street food. It's hard work! You think it's easy? Men do big things, they can't do meticulous things." Trung, a crab meat noodle seller, adds, "Men don't like doing [petty] business, I mean, dealing with small things, such as a few thousand [*đồng*] for one bowl of noodle." (At the time of writing there were 20,000 Vietnam *đồng* [VND] to US$1.) The implication is that something as low value and fiddly as food selling is women's work, and such a perception is reinforced by the lack of men selling food in the Vô Danh street market.

RELATIONSHIPS WITH AUTHORITIES IN AN INFORMAL STREET MARKET

Vô Danh is an informal street market in Ha Noi's Old Quarter, which means it is not legal. The vendors who work there know they face possible fines and the confiscation of their goods. Fear makes some of them hyper-vigilant and anxious. As Thien explained, "I first started selling vegetables and then switched to fish because vegetables are too heavy. Fish is lighter and I can run faster when the police come." Tue, a tea seller, said, "The market used to be bigger with more people selling but the police presence is bigger too, and the market has shrunk in size." There is also a disparaging discourse that circulates in the society about vendors from outside Ha Noi, which is the case for almost all itinerant street sellers who carry their goods using a yoke and two baskets. Such vendors identify themselves as among the poorest of their kind, lacking a stable location, the financial resources required to buy and store produce, and the bicycle or cart which could be used to transport it.

Although we interviewed no police or security force members, because of the issues around (il)legality, we witnessed a number of encounters between law enforcers and street sellers. Some interactions seemed reasonably light-hearted, with officials representing the local ward (the Vietnamese unit of local government, comprizing a number of Ha Noi city blocks) chasing vendors away on foot. In those cases, the officers ceased pursuit after about 20 yards and the street vendors returned as soon as officials left the scene. Sau, a pork sausage seller, told us, "The [local] police say I'm not allowed to use the pavement but I generally hide by the tree, and hope they will ignore me." Vendors are not overly fearful of this level of authority. Matters do not always resolve so benignly, however.

The state police (bright green uniforms) and ward-level security forces (dark green uniforms) are more likely to act punitively. On several occasions we saw them swoop into the street in a canvas-covered truck containing some half dozen men who would jump out of the back and confiscate baskets of produce from itinerant vendors. Sometimes the truck was followed by a motorcycle backup. About half the time, the law enforcers idled the truck part way along the street but did not stop, apparently satisfied with the consternation they caused. At other times they chased the vendors, sometimes using the motorcycle to overtake a fleeing vendor. I saw baskets of produce confiscated from poorer vendors, women with yokes selling sugar cane and pineapples. The goods were put into the truck and driven off.

During one such happening, Ngoc, a woman who sells sugarcane, was furious rather than cowed. She chased the police van as it drove away, shouting, "Thieves!" She later told us she'd given the police a 50,000 VND 'donation' only the previous day. This was equivalent to more than two days' earnings, while the loss of the basket and its contents would wipe out at least another day. Such vendors might hope to make US$20 profit in a month (Jensen, 2012). It transpired, however, that Ngoc actively protected herself against total calamity. She paid a local shopkeeper to hide stock for her and was selling again within minutes of the police leaving. It seemed to me that she sold more sugar cane following the police raid, as if local people wished to support her. When Ngoc eventually moved on from the street an hour or so later she creatively used an improvised wire mesh circle with a lining of plastic bags in place of the missing basket on her yoke.

Even established shop owners fear law enforcers. On a separate occasion we watched uniformed officers fine first a fruit shop owner and then a shop selling *bánh mỳ*, a takeaway bread roll dish which is served with a range of cooked fillings (see Green & Nguyen, 2013). In each of the shops the officers asked for and were given a seat and a drink, and then, once comfortable, they proceeded to write out a charge for obstructing the pavement. It was impossible to know why these shops had been selected when every shop in the street was either displaying its own goods on the pavement or had leased that space to a street vendor. Van, my research assistant/interpreter, told me that the uncertainty of these events and the randomness of good and bad luck mean that street traders are superstitious and are always on the lookout for signs of good or bad fortune. This is particularly the case early in the day, at the beginning of a week, and at the start of each lunar month and lunar year. It is rude to bargain too fiercely at such times because more is at stake than the sale itself, and a fiercely contested sale can be construed as the harbinger of a bad day, week, month, or year for the vendor.

CHANGING ECONOMIC IMPERATIVES

Street selling first received significant regulatory attention in Vietnam in 1989. Since then, Vietnamese authorities have focused upon street food hygiene, predominantly in terms of educating vendors and licensing them (Bhowmik, 2012, p. 37). Such a food safety regime has some positive impact on the vendors who have access to a secure shop-space, with a lockable area, a water supply, and washable work surfaces, but these people are well-off compared with itinerant sellers who have to negotiate with a range of street residents to use clean water and storage.

Regular customers watch the various compromises around hygiene and cleanliness and decide what to buy from whom, according to a range of personal criteria. Mai had only recently set up her goose noodle soup shop in the market street. "Where I used to work, I was able to sell 15 to 16 geese a day. Here I only sell 10 geese a day, that's because of the change of location. They don't know me as well here, but it will gradually get better, and I'll get more customers as people get to know me." Doan, a meat seller, learns to buy the kind of meat her customers want. She told us, "I start work at 5:00 a.m. and I buy meat from a bigger market [. . .] I buy from the same people every day, although I don't know where the meat comes from. What I do is I pick the best looking meat for that day." Chi, who sells chickens, guarantees the quality of her product by selecting her wares when they are still alive:

> They are killed at the same place that I buy the chickens, by a person who kills them for me. That's the way I know they're fresh [. . .] I sell free range chicken. It tastes better, even though it's more expensive than factory raised chickens. My customers are upper income and they want good food. If people want big quantities of my chickens, they call me in advance. Usually I only buy a few chickens [. . .], only a few people can afford this kind of free range chicken.

Fish seller Thien told us that

> Selling fish is easier than selling meat, as long as you keep them alive. Most customers keep coming back to the meat seller they already know, but for fish they only care about freshness [. . .] It's easier to sell meat, because people can use it for many more flexible dishes. But it's easier to set up store with fish, because customers are pickier with meat.

Given the importance of selling fish fresh, successful vendors improvise ways of keeping them alive. Thien's fish are kept in a tank in the storage compartment under her motorcycle seat:

> [I]t's kept aerated with an air bubbler. The tank is kept separate from the [motorcycle] battery and the engine [. . .] I store them between my feet when I drive along. I have to carry the water with me all the way from home and sometimes I have ice in the summer.

The inventiveness of some vendors, and the initiative they use in building links with their customers, indicate their entrepreneurship and demonstrate that some women have prospered in their self-employed food vendor role in the Vietnamese economy.

Many older street sellers speak nostalgically of the days when Vietnam was more closely aligned with the Soviet Union, before "the period of reform *(doi moi)* [. . .] (1986–present)" (Watts, 1998, p. 426). These changes were exacerbated in the popular imagination by the disintegration of the USSR after 1989. Women often locate their role as street vendors as an outcome of the closure of many state-owned enterprises over the past quarter century. Tue, the tea seller, said, "I used to work in a garment factory but the factory closed down, because it had been funded by the Soviet Union. I was laid off." Ma, the meat seller, used to be "a construction worker for the government and [my] children were sent to school for free. It's harder now, because that was in the old system. Now people have to find their own way of making a living." There is an apparent nostalgia for the benefits of the old ways under a system of state-sponsored employment. But the speakers generally represent women who face challenging financial circumstances and who are located in north Vietnam, Ho Chi Minh's power base during 'the American War.' The south had fewer state-run enterprises.

Street vendors say they work to cover basic necessities and in the hope that their children can stay in school, finish their education, and have better employment choices. Hoa, who has been selling sticky rice for a year, said, "Office life is a lot easier, but you need a degree." Chi is resigned about the economic choices she made when she started her family:

> I used to work as a kindergarten teacher [. . . until] I was pregnant, because the income as a teacher wasn't enough to keep us all. Now teachers earn better income than street sellers, but at that time you could earn more in the market.

Sau also faced employment challenges when she started her family:

> My husband died 20 years ago. He died when my second son was 2 years old. But I already worked here at that point. I used to work as a mechanic in a state-owned company, but [. . .] I was laid off because I had young children and a husband.

The pressure to keep earning continues even after a woman's dependents have made their own way in the world. Doan, a meat vendor, told us that she sells mainly "to pay for lunch and breakfast." She continued: "I no longer have to worry about my husband and children. The children are grown up and have their own families. I only have to look after myself [. . .] since I retired." She works eight hours a day, from 5:00 a.m. until about 1:00 p.m. By that time, most people have bought their fresh ingredients for the day's cooking, both lunch and dinner.

THE CITY AND COUNTRY

Vendors in the Vô Danh informal street market include residents who live along the street, people who live elsewhere in the city but work in the street, and itinerant street sellers who divide their time between living in the city and brief visits to their homes and families in the country. Both Hoa and Tue live on the street itself. Hoa lives with her husband and baby in her mother-in-law's home, as is traditional in Vietnamese society. She sells sticky rice in the morning so that she can spend the afternoon with her family. She works Monday to Saturday inclusive, starting her cooking at 4:00 a.m.

and usually closing her stall sometime around noon, but definitely before 2:00 p.m. After closing up, Hoa clears away, washes utensils, and buys and prepares some of the food for the next day's sales. Her stall is supported by a range of deliveries, some of which are sourced from beyond Ha Noi:

> One person delivers sausages and another person delivers pâté and other meat, a third delivers bread, all by 6:00 a.m. [. . .] The people who were making this food [when I moved in] refused to tell me who the suppliers were, so I had to approach the deliverers directly. That's how I got started about a year ago.

Tue is also up at 4:00 a.m. She used to sell tea only in the evening but her student-aged clientele asked her to do breakfasts too, so she started making them sticky rice: "When it rains, I can't work [. . .] My customers call me on the phone and say they're sad because they've got nowhere to sit." While we were interviewing, a customer asked Tue if he could borrow her phone so he could call a friend and tell him where he was. Tue, who had his friend's number on her phone, dialed it, and handed it over to him. Her stall is clearly an important part of some of her customers' lives. "They all call me mom. I don't take advantage of anyone. They know I'm kind, and they like me. They look for me when I'm not here and call me up to check if I'm OK." Tue sleeps in two shifts: a siesta at noon, a couple of hours at night. She has worked in the market since the early 1990s and is on the street for about six hours per day, between 6:15 a.m. and 8:00 a.m. (selling sticky rice) and from 7:00 p.m. to 11:00 p.m. (selling tea). She says the rest of her time is taken with shopping, cooking, and clearing away her products, or in doing her own housework.

Doan and Chi both live about 3 km away, close enough to go home for a rest in the middle of the day, sometimes from 10:30 a.m. to 3:00 p.m. Mai lives closer, on a major thoroughfare nearby. Sau, the pork sausage vendor, lives near enough to call up further supplies from her mother-in-law and sister-in-law if her meat runs low. That way she limits the amount of food waiting to be sold a small quantity at a time. Thien's home is in the country, a 90-minute bicycle ride from Ha Noi. Over the past couple of decades she has made enough money to upgrade from selling vegetables to selling fish. Now she can afford a motorbike, bringing her journey time down to 40 minutes each way, and allowing her the means to keep her fish alive for longer.

All these women are comparatively advantaged. This is not the case with sugar bun seller Phuong. Like Ngoc, whose sugar cane and basket were confiscated by police, Phuong lives too far away from Ha Noi to return home at night. Instead, vendors who use a yoke and baskets tend to work a 10–12 day fortnight, returning home for two to four days every two weeks or so. When in the city, Phuong lives in a hostel with her sister and 40 other women who also sell buns. These women typically sleep 10 in a room, paying about US$0.35 (7,000 VND) a night to do so (Jensen, 2012).

Phuong sells four varieties of sugar bun, three of which she makes herself. She buys the fourth from a specialist baker who supplies all the sellers in the sugar bun hostel. Sugar bun baking starts at 2:00 a.m., and Phuong is usually selling by 5:30 a.m. She takes a nap at noon but can work as late as 10:00 p.m. to 11:00 p.m., hoping to sell all her goods before they become stale and inedible. A widow, Phuong is better off than the many Vietnamese farmers whose lands have been compulsorily acquired with little compensation as a result of urban expansion (Brummitt, 2013). Phuong has access to a small area of arable land but works in the city to clear her debts. "I would like to stay home to work on the farm. I work in the city when the farm is able to run by itself, because it gives me extra income." Phuong told us that last year was a bad one, as she had many debts to clear: "I had to spend 11 million dong on a computer for [my daughter's] IT courses. [. . .] As soon as I had taken out a loan of money to pay for the computer, my father died. And then I had to take out a loan of money to pay for the funeral." Close to the poverty line, life becomes precarious. She told us that when her well-educated daughter marries the daughter will live with her husband's family and help with their finances, rather than with Phuong's, but she seemed to take comfort in this view of her daughter's future prosperity.

The debts discussed by the women in the research were often related to personal obligations which fell upon a daughter in the absence of a son, or upon a mother who had been widowed or abandoned, or whose own husband was ill or unable to work. Some of the nonworking men had 'missing papers,' which meant they don't receive a monthly allowance for fighting against South Vietnam in the American War. A husband's missing papers was one of the many reasons given by women for participating in the economic activity of the informal street market.

DISCUSSION

This chapter charts an encounter with a community of women working in the Vô Danh street market in Ha Noi, Vietnam. These vendors in the informal street market rely upon their understanding of their customers' preferences to cover their costs and earn the money required for survival. Many say they have no safety net other than their own enterprise. Older women cannot necessarily rely on the support of adult children unless they have a son, because daughters owe allegiance to their husband's family. Gibson (2010) argues that it is these "encounters between producers and consumers that make more transparent the politics of capitalism" (p. 522), but they also make visible the workings of gender in the construction of poverty.

Informal street markets provide an economic location for workers who are excluded from regulated employment or who choose a space in which they can be entrepreneurial. Vô Danh helps illuminate some less-acknowledged aspects of Vietnam's economy. I had anticipated that the chapter would focus on food and the treatment of animals, because my first encounter had been at the level of a tourist gaze (Urry, 1990) inflected through veganism. Instead, it was through my gender that I discovered a "sense of connectedness [which] is temporary, but [which] can reflect a deeper reactivation of concerns" (Spinosa, Flores, & Dreyfus, 1997, p. 136). It is difficult to talk to the women of Vô Danh, most of whom work for the welfare of family members as well as themselves, without becoming aware of the disparities in economic and political opportunity and different levels of social support.

A plum seller, Ly, asked me why we were doing the research if it wasn't going to make a difference to her life and the lives of other itinerant vendors. She interviewed me as to the motivations and interest in her experiences. She turned her gaze on me, specifically. I could only reply that I believed there was value in telling her story to a different audience, and it was part of a process of building awareness of the problems she faces, of influencing the response of nongovernmental organizations and everyday tourists. Such awareness might have eventual impacts upon Vietnamese and Ha Noi policy makers, and inform a dialogue which could open up avenues for possible change. There is a growing interest in the lives of hawkers and other itinerant street vendors, as evidenced in a recent exhibition in the Vietnamese Women's Museum (Jensen, 2012), and this chapter helps—however modestly—to promote discussions around their lives, the repression they face, and the financial debts they feel bound to incur.

These experiences of speaking to women selling food in the Vô Danh market provided me with a means of "reflecting and indeed producing interethnic commensalities and disjunctures" (Wise, 2011, p. 84). Privileged access to the women's perspectives, through the work of Van acting as interpreter/translator, made it possible for me to find common threads with the lives of many of these women, even as our discussions also highlighted differences. I had expected that the tourist gaze would be an important aspect of our discussions, yet it was the "inspecting gaze" (Foucault, 1980, p. 155) of the authorities which had the most impact upon the research participants.

Many of the street vendors we spoke to felt they had no choice but to work in the informal economy. Their work was grueling and protracted—cold in winter, hot and sticky in summer; with significant risk and no security—but some told us it was the only opportunity open to them. Their participation in an informal street market pits them against various levels of authority whose aim remains to control the women's activity and to restore the streets to order. The Vô Danh street market represents a wide range of interconnected social and economic ties which appear to have little relevance to people from outside Vietnam. Indicating that this is not the case, Brenner (2009) notes that "geographical difference no longer represents the specialization of particularity. It instead demarcates the distinctive positionality of any given space within an evolving, worldwide grid of interdependencies" (p. 29). Tourists are implicated within these interdependencies, not least because they tend to engage in metropolitan-based activities, encountering country people and produce as part of the extended system which supports life in the city. This study of how some women have to find their own way of making a living in an informal street market in Ha Noi's Old Quarter may modestly illuminate the development and realization of such interdependencies.

ACKNOWLEDGEMENTS

I am indebted to the work of my research assistant and interpreter Ms Nguyen Hong Van, and to the Faculty of Education and Arts, Edith Cowan University, for authorizing Study Leave in Semester 1, 2013, during which I conducted the fieldwork for this chapter.

REFERENCES

Bhowmik, S. (2012). Street vendors in Asia: Survey of research. In S. Bhowmik (Ed.), *Street vendors in the global urban economy* (pp. 20–45). New Delhi: Routledge.

Brenner, N. (2009). A thousand leaves: Notes on the geographies of uneven spatial development. In R. Keil & R. Mahon (Eds.), *Leviathan undone? Towards a political economy of scale* (pp. 27–49). Vancouver: University of British Columbia Press.

Brummitt, C. (2013, January 31). In Vietnam, rage growing over loss of land rights. *Associated Press*. Retrieved from http://bigstory.ap.org/article/vietnam-rage-growing-over-loss-land-rights

Douglas, M. (1966). *Purity and danger: An analysis of concepts of pollution and taboo*. London: Routledge & Kegan Paul.

Foucault, M. (1980). The eye of power. In C. Gordon (Ed.), *Power/knowledge: Selected interviews and other writings 1972–1977 by Michel Foucault* (pp. 146–165). Sussex, UK: Harvester Press.

Gibson, C. (2010). Geographies of tourism: (Un)ethical encounters. *Progress in Human Geography, 34*(4), 521–527.

Green, L., Costello, L., & Dare, J. (2010). Veganism, health expectancy, and the communication of sustainability. *Australian Journal of Communication, 37*(3), 51–72.

Green, L., & Nguyen, V. H. (2013). Cooking from life: The real recipe for street food in Ha Noi. *M/C Journal, 16*(3). Retrieved from http://journal.media-culture.org.au/index.php/mcjournal/article/viewArticle/654

Jensen, R. (2012). *Street vendors* [DVD of three films: "Their voices", "Thuy's story", and "Loi's story"]. Ha Noi: Vietnamese Women's Museum.

Singer, P., & Mason, J. (2006). *The ethics of what we eat: Why our food choices matter*. New York: Rodale.

Spinosa, C., Flores, F., & Dreyfus, H. (1997). *Disclosing new worlds: Entrepreneurship, democratic action, and the cultivation of solidarity*. Cambridge, MA: MIT Press.

Tantiwiramanond, D. (2004). *Changing gender relations and women in micro enterprises: The street vendors of Hanoi: A research report*.

Urry, J. (1990). *The tourist gaze*. London: Sage.

Urry, J. (1992). The tourist gaze "revisited". *American Behavioral Scientist, 36*(2), 172–186.

Watts, M. (1998). Recombinant capitalism: State, de-collectivisation and the agrarian question in Vietnam. In J. Pickles & A. Smith (Eds.), *Theorizing transition: The political economy of post-communist transformations* (pp. 425–478). London: Routledge.

Wise, A. (2011). Moving food: Gustatory commensality and disjuncture in everyday multiculturalism. *New Formations, 74,* 82–107.

3 On Being and Becoming in Melbourne's Marketplaces

Maša Mikola

Globalization may have reduced the significance of local marketplaces and replaced them with globalized spaces of consumption such as shopping malls, yet markets are still important spheres of exchange and interaction. Their continued popularity and cultural significance show us that they are not fully confined to the logic of consumption in late capitalism (Adorno, 2000). This chapter considers two markets in Melbourne—Queen Victoria Market and Footscray Market—as a case study on how marketplaces retain their role as places of community interaction. It focuses on their spatial and social location, their connection to the world 'outside' (the street, the suburb), the languages used in the marketplace, and how these are framed by political and cultural discourses of multiculturalism in Australia. It draws on philosophers Gilles Deleuze and Félix Guattari's (1986; 2004) discussion of minor and major literature, and transposes those ideas from the literary field to that of the market. The analysis of the interplay between minor and major literatures is used as a model to situate the more informal and peripheral Footscray Market in relation to Queen Victoria Market, where the experience could be characterized as more regulated and managed, spatially and affectively; therefore, belonging to the formal (cultural) economy of the city. The subversion of the major by the minor, and the subsumption of the minor by the major, as analysed by Deleuze and Guattari (in relation to the writing of Franz Kafka), are processes which are explored within the context of Melbourne's marketplaces. In addition, the chapter mobilizes data collected from ethnographic research at and around Footscray and Queen Victoria markets from 2006 to 2009.

LOCATING THE MARKETS

Queen Victoria Market and Footscray Market may not seem to have much in common, apart from the fact that they are both markets, because they belong to different narratives of Melbourne's history and the national history of Australia. The former has subsumed and edified its migrant history within a formal, dominant national history, while the latter still sits at the periphery of this narrative, both questioning it and waiting to be included.

Queen Victoria Market is located on the north-western boundary of Melbourne's Central Business District (CBD) and is managed by the Melbourne City Council. It is the only surviving 19th-century market in the centre of Melbourne, and is listed on the Victorian Heritage Register. It was built on land formerly occupied by the city's main cemetery. When the cemetery was moved to another location in 1854, the land was granted to the Melbourne Town Council in order for them to build a market. Most bodies were exhumed, although the unmarked graves—mostly of ex-convicts, the working class, Aboriginals, and others who could not afford a proper burial—were left on-site. Throughout the 20th century, Queen Victoria Market was a central location for employment for many postwar migrants in Melbourne. Greek, Italian, and Polish families established shops at the market, with subsequent waves of migration expanding the diversity of this milieu, to the extent that Queen Victoria Market is today considered an icon of multiculturalism in Melbourne. It is promoted as such by the Melbourne City Council, who position the market as "much more than just Melbourne's shopping mecca—it's a historic landmark, a tourist attraction and a Melburnian institution" (Melbourne City Council, 2013, para. 1).

Prior to European settlement, the current site of Footscray Market was an important meeting place for the tribes of Yalukit-willam, Marin-balluk, and Wurundjeri (Lack, 1991). Expanding industry in the second half of the 19th century, and Footscray's close proximity to the city's docks, saw the construction of a railway bridge over the Maribyrnong River, followed by a tramline in the 1920s (Lack, 1991). Tram and train connections made the suburb more accessible to the central districts of Melbourne, five kilometres away. Despite this proximity, Footscray's social and spatial marginality was derived from its location in the city's industrial West, which was both an undesirable place to live and a place to house 'undesirables.' In the 1950s and 1960s, Italian, Greek, Macedonian, Bosnian, and Croatian migrants settled there. As a direct consequence of the conclusion of the Vietnam War in 1975, 54,000 Vietnamese people were resettled in Australia between 1977 and 1982 (Stevens, 2012, p. 526), with many settled in reception centres in areas such as Footscray, "which had not previously been characterized by Asian settlement" (Collins, 1995, p. 376). The arrival of Vietnamese refugees in Footscray coincided with the termination of the racially restrictive White Australia Policy in 1973 and Australia's ideo-geographical reorientation as 'part of Asia' (Jupp, 1995; Lovell, 2007; Poynting & Mason, 2008).

MULTICULTURALISM IN THE MARKETS

The histories of Queen Victoria and Footscray markets are part of a narrative of migration that has defined Australia since European settlement in the late 18th century. The White Australia Policy was instituted with the Immigration Restriction Act, the first act that passed through the parliament of the Federation of Australia in 1901. The policy favoured white British immigrants, and was discriminatory against other races and nationalities, particularly the Chinese. The policy was in place until 1973, when it was officially dismantled by the socially progressive Whitlam Labor Government.

Despite the strong political and social push at the beginning of the 20th century to create an ethnically homogenous nation (Jupp, 1998, p. 132), this picture changed radically in the second half of the 20th century. When multiculturalism was adopted as government policy, Australia promoted and perceived itself as an ethno-culturally diverse place. Since its introduction, however, multicultural policy has passed through different stages. The initial focus on the concept of 'cultural pluralism,' made up of distinct ethnic communities of largely unskilled labourers during the first decade, shifted to a focus on skilled migrant workers during the 1990s, along with the alignment of multiculturalism as 'identity politics' (Jayasuria, 2008; Hage, 2003). During the socially conservative Liberal government of John Howard (1996–2007) many policies supporting multiculturalism as a discourse were abandoned. In 2011, bipartisan political support for multiculturalism was announced again (Colic-Peisker, 2011), with renewed focus on permanent and temporary skilled migration.

Promotion of multilingualism was an important part of the politics of multiculturalism in its formative years. However, there has always been an assimilative trend present, with social and cultural institutions working to position English as the sole official language used in Australia (Ang, Hawkins, & Dabboussy, 2008). Current immigration schemes mandate prospective migrants fulfil requirements for English proficiency, irrespective of their reason for migrating, which has the effect of amplifying assimilationist linguistic policies.

Eighty-one per cent of Australians speak only English at home (ABS, 2013), whereas Mandarin, the most common language after English, is spoken by 1.7 per cent of the total population. Use of languages other than English, in both Footscray and the City of Melbourne, is, however, much higher than the national average. My ethnographic research and analysis of languages spoken at the markets showed that Italian, Greek, and Polish are the main languages, other than English, at Queen Victoria Market. This is not reflective of the larger City of Melbourne, however, where the most common non-English languages are Mandarin, Cantonese, and Indonesian (ABS, 2011b). In Footscray Market, on the other hand, Vietnamese is the dominant language. Australian Bureau of Statistics (2011a) figures confirm this picture. In 2011 in Footscray 51.9 per cent of people spoke at least one other language besides English at home. Of this group, 11.8 per cent spoke Vietnamese.

There is a difference in the way nonofficial Australian languages are used in the two markets. At Queen Victoria Market, the use of Italian, Greek, and Polish performatively attends to the market's multicultural brand rather than an everyday communicative function, as these are not, as the Australian

Bureau of Statistics figures show, the languages used by the people who live in Melbourne's CBD. The languages and voices I heard at Footscray Market only rarely produced this performative branding. Cultural difference in Footscray is informal. It is not something exotic you cloak yourself in on a Saturday morning to go shopping for groceries. A Scottish fish trader and his entirely non-Vietnamese staff learned to speak Vietnamese in order to better communicate with customers. As he related:

> Ninety percent of our clientele or . . . 80% . . . because we've got a lot of Chinese as well . . . 80% are Vietnamese. When we're dealing with the public, we talk mostly in Vietnamese and we have no Vietnamese worker behind the counter. Not for any reason . . . there's no reason why we don't employ them. We've all been together, we started together and we've stayed together. Two of them went to school to learn Vietnamese, the rest of us just learn it through working. We've got Aussies behind the counter, we've got Indians behind the counter . . . and we can speak Asian and an old lady, 80 years of age being Vietnamese, can't speak a word of English, she still comes up to our shop, be confident by herself, to buy fish, get the right change back. So, it's a huge benefit.

At Footscray Market 'foreign' language is not really foreign, because it is a language that is also used outside the market. Language connects the market to Footscray; its shops, restaurants, and people, and creates a spatial informality that links the market to the life of the street, and, to a certain degree, is a recreation of the informality predominant in street markets in Vietnam.

Because the market is frequented by Vietnamese and non-Vietnamese shoppers, produce signs, prices, and other communication are often in both Vietnamese and English, and retailers switch between languages. Sometimes, however, the linguistic barriers between traders and shoppers are too pronounced, and body language is used. The language used is constantly being renegotiated and is therefore more dialogical. Dialogical and relational aspects of people's interaction across cultures (Bakhtin, 1986)—body language besides words—are regularly forgotten in multicultural policies.

Footscray Market is an internally heterogeneous community, or what Sakai and Solomon (2006) call a "non-aggregate community," where "you are always confronted, so to speak, with foreigners in your enunciation" (pp. 8–9). By contrast, the notion of a multicultural community in public policy often dwells, paradoxically, in the idea of unified language and values. As Zygmunt Bauman (2001) and Fran Tonkiss (2003) argue, there is no acknowledgement of the fact that community means different things to different people. For some, it is a concept and instrument for unity; for others, it connotes a sense of imprisonment. In relation to multicultural spaces and practices, a preoccupation with the idea of a unified community means that the designers of the spaces that are planned to be multicultural forget about leaving the spaces *blank* or *silent*; they forget about allowing for the translation between old and new, colonized and colonizers, settlers and migrants.

The range of languages used at Footscray Market includes Vietnamese, Macedonian, Serbian, Chinese, Australian English—although in the confluence of intersecting sounds and identities, the broken speech is *none* of these languages. English, or Vietnamese, or Chinese, may be used in a unique way, where words are left out or interspersed between languages. Alice Pung, a second-generation migrant Australian writer who spent a lot of time at Footscray Market, where one of her parents worked, writes about this in her memoir of growing up in Melbourne's western suburbs:

> This is the suburb where words like and, at and of are redundant, where full sentences are not necessary. "Two kilos dis. Give me seven dat." If you were to ask politely, "Would you please be so kind as to give me a half-kilo of the Lady Fingers?" the shop owner might not understand you (. . .) To communicate, my father realizes, does not merely mean the strumming and humming of vocal cords, but much movement of hands and contortion of face.
>
> (2006, p. 2)

RESISTANCE IN THE MARKETPLACE

The two markets act as two narratives—one minor, the other major—that are variables of the same language, or "two different treatments of the same language" (Deleuze & Guattari, 2004, p. 114). Queen Victoria Market is where minor languages appear in a located and bounded manner. They serve a performative function within a market that is a product of its historical connection to Melbourne's colonial past, its central urban location, and its function as heritage site and tourist attraction. Consequently, Queen Victoria Market belongs to a dominant narrative of market spaces.

With regard to major and minor, Deleuze and Guattari (2004) differentiate between language and literature. Minor language is a language of the minority and is imbued with "the power of variation," whereas major language can be defined by "the power of constants" (p. 112). A minor language can be made major by its users, through the capacity to force official recognition. Minor literature, on the other hand, "doesn't come from a minor language; it is rather that which a minority constructs within a major language" (Deleuze & Guattari, 1986, p. 16). What characterizes minor literature, in their view, is the use of a major language by a minority that therefore 'deterritorializes' the major language, thus subverting it. Deleuze and Guattari cite the example of Kafka, a Czech Jew writing in German. Prague German is "a deterritorialized language, appropriate for strange and minor uses" (1986, p. 16); it is both German and not-German.

Transposing this idea to marketplaces may seem as strange as Kafka's literature itself, but it is in these spaces of exchange that language assimilation, or lack thereof, becomes a language unto itself. If we understand Footscray Market as a form of minor literature (as opposed to the minor languages spoken there), we begin to see how, through the performative aspects of the place (the *Gestalt* of its particular verbal and nonverbal languages, the characteristics of its surrounding streets and suburb, its informality), it becomes a minor marketplace in relation to Queen Victoria Market. We open up the possibility of a dialogue in a shared and major language, which, at the same time, challenges, subverts, and deterritorializes that major language. A minor literature is inherently political, as it deals with the definitions and boundary conditions of community in dialogue with the identity of a city, or a nation. Deleuze and Guattari (1986) ask:

How many people today live in a language that is not their own? Or no longer, or not yet, even know their own and know poorly the major language that they are forced to serve? This is (. . .) the problem of minorities, the problem of a minor literature, but also a problem for all of us: how to tear a minor literature away from its own language, allowing it to challenge the [major] language and making it follow a sober revolutionary path?

(p. 19)

If resistance or opposition is one function of difference or of living with difference, then the place of difference needs to be at least partially unpredictable. The question here is whether unpredictability in the marketplace can pose any kind of resistance to a dominant narrative into which the marketplace might be tied. And, if it does, how is this achieved?

For Deleuze and Guattari (2004), what actually separates the major and the minor is the sense of becoming. "Becoming-minoritarian" is an ethical position, or an opposition, or a state that announces a new way of being through a process of change and movement. They say that the minor is the "becoming of everybody," while the major is "always Nobody" (p. 117). There are minor languages present in both markets, yet that is not to say that they are in a state of becoming at both markets. Queen Victoria Market does not challenge the major language—standard Australian English—from within, nor does it challenge the major literature, which, in this case, is white Australia's construction and regulation of the national discourse of multiculturalism. At Queen Victoria Market languages exist in a kind of stasis to be preserved, defended, and displayed. The market has been subsumed and formalized within the nation's myths so as to be presented as a historical artefact and tourist attraction. Migrant languages are present, but they are not in a state of 'becoming,' and instead have been fully incorporated into the major literature of the market.

On the other hand, change and movement are inscribed not only in Footscray Market, but also the experience of the suburb of Footscray as the market's location. As one of the interviewees observed:

> It never felt like there was an official entrance at the Footscray Market, because what we were doing outside was quite similar to what we were doing inside. So, opposite Footscray Market, I think it's Barkly Street. Opposite that, there's a big kind of grocery store called Dat Sing, and we would go there, depending on where we parked, go there first or go there last. Because Footscray wasn't just the market for us. We would go to the Footscray Market as one of the stops. Dat Sing is another stop and then also to the back part where BI-LO is. It's called little Saigon now. BI-LO was a name for the entire little complex there, even though you might have not used BI-LO in particular. So, our kind of Footscray experience would be really a big giant market experience. It wasn't just going into a market, it was going to several.

Footscray Market is effervescent and dynamic, as is the suburb in which it is situated. In contrast to the major, which is the formality of regulated space and flows in the market and the streets, the intersection of bodies and communications taking place is informal and minor.

Informality characterizes the practices of infiltration (Dovey & King, 2011, p. 12) at work at Footscray Market. In an architectural sense, the space here is more bounded and enclosed than Queen Victoria Market, but its boundaries are constantly traversed and renegotiated. Several small eateries line the external boundaries of the market through which shoppers, workers, and visitors can pass in order to enter the market or the street. The dynamic of the market is the dynamic of the street outside the market, because those boundaries are porous and are rarely sensed. Queen Victoria Market, by contrast, appears more open and spacious, yet its boundaries are more formalized and stable. Different sections at the market do not flow into one another, and the barriers between sections are clearer:

> The Victoria Market is definitely . . . well, apart from the fact that it is huger, there's that huge section of general merchandise, and it's all in lanes, in blocks and columns. Big square section of that and then the big square section of lanes and vegetables, and then you've got the food court down the bottom, meat section all in lanes, all going the same way actually, they don't seem to go across at all. Then organic produce has a big section as well, so you know exactly where you're getting, where you're going to get something.

The space of Queen Victoria market is predictable; "you know exactly where you're getting." The space is governed by formally established practices of selling and buying, and there is not much communication between the market and the street, despite it being considered an open market.

Seen as a 'minor literature' in relation to Queen Victoria Market, Footscray Market challenges the politics of mainstream, top-down multiculturalism. If the 'major literature' is framed generally as the politics of national identity, then the minor and the major not only form a dialogue that involves both the subversion of the major by the minor, but also the inextricable process of the minor moving towards and 'becoming' subsumed by the major. The change in Footscray over the last few years, as the suburb has become more gentrified and as the demographics have shifted to incorporate young families and professional business people, can be seen as a further example of how the status of the minor is always in the process of becoming a part of the major, whilst resisting, challenging and changing it. As Annie Brisset (2003) observes, "the ethics of difference (. . .) may well coincide with an ethics of resistance" (p. 18).

REFERENCES

Adorno, T. (2000). *Problems of moral philosophy*. Cambridge, UK: Polity Press.

Ang, I., Hawkins, G., & Dabboussy, L. (2008). *The SBS story: The challenge of cultural diversity*. Sydney: University of New South Wales Press.

Australian Bureau of Statistics (ABS) (2011a). *2011 Census QuickStats: Footscray*. Retrieved from www.censusdata.abs.gov.au/census_services/getproduct/census/2011/quickstat/SSC20496

Australian Bureau of Statistics (ABS) (2011b). *2011 Census QuickStats: Melbourne City*. Retrieved from www.censusdata.abs.gov.au/census_services/getproduct/census/2011/quickstat/20604

Australian Bureau of Statistics (ABS) (2013). *Cultural diversity in Australia. Reflecting a nation: Stories from a 2011 census*. Retrieved from www.abs.gov.au/ausstats/abs@.nsf/Lookup/2071.0main+features902012–2013

Bakhtin, M.M. (1986). *Speech genres and other late essays*. Austin: University of Texas Press.

Bauman, Z. (2001). *Community: Seeking safety in an insecure world*. Cambridge, UK: Polity.

Brisset, A. (2003). Alterity in translation: An overview of theories and practices. In S. Petrilli (Ed.), *Translation, translation* (pp. 102–132). Amsterdam: Rodopi.

Colic-Peisker, V. (2011). A new era in Australian multiculturalism? From working-class "ethnics" to a "multicultural middle-class". *International Migration Review, 45*(3), 562–587.

Collins, J. (1995). Asian migration to Australia. In R. Cohen (Ed.), *The Cambridge survey of world migration* (pp. 376–378). Cambridge: Cambridge University Press.

Deleuze, G., & Guattari, F. ([1975] 1986). *Kafka: Toward a minor literature* (D. Polan, Trans.). *Theory and history of literature* (vol. 30). Minneapolis: University of Minnesota Press.

Deleuze, G., & Guattari, F. (2004). *A thousand plateaus: Capitalism and schizophrenia* (B. Massumi, Trans). London: Continuum.

Dovey, K., & King, R. (2011). Forms of informality: Morphology and visibility of informal settlements. *Built Environment, 37*(1), 11–29.

Hage, G. (2003). *Against paranoid nationalism: Searching for hope in a shrinking society*. Annandale: Pluto Press Australia.

Jayasuria, L. (2008). Australian multiculturalism reframed. *New Critic, 8*. Retrieved from www.ias.uwa.edu.au/new-critic/eight/?a=87787

Jupp, J. (1995). From "White Australia" to "part of Asia": Recent shifts in Australian immigration policy towards the region. *International Migration Review, 29*(1), 207–228.

Jupp, J. (1998). *Immigration*. Cambridge: Cambridge University Press.

Lack, J. (1991). *A history of Footscray*. North Melbourne: Hargreen Publishing and the City of Footscray.

Lovell, M. (2007, September). *Settler colonialism, multiculturalism and the politics of postcolonial identity*. Paper presented at Australasian Political Studies Association Conference, Monash University, Melbourne.

Melbourne City Council (2013). *Queen Victoria Market*. Retrieved from www.thatsmelbourne.com.au/Placestogo/MelbourneLandmarks/Historic/Pages/4144.aspx

Poynting, S., & Mason, V. (2008). The new integrationism, the state and Islamophobia: Retreat from multiculturalism in Australia. *International Journal of Law, Crime and Justice, 36*, 230–246.

Pung, A. (2006). *Unpolished gem*. Melbourne: Black.

Sakai, N., & Solomon, J. (2006). Introduction: Addressing the multitude of foreigners, echoing Foucault. In N. Sakai & J. Solomon (Eds.), *Translation, biopolitics, colonial difference* (pp. 1–39). Hong Kong: Hong Kong University Press.

Stevens, R. (2012). Political debates on asylum seekers during the Fraser government, 1977–1982. *Australian Journal of Politics and History, 58*(4), 526–541.

Tonkiss, F. (2003). The ethics of indifference: Community and solitude in the city. *International Journal of Cultural Studies, 6*(3), 297–311.

4 Migrants in Informal Urban Street Markets

Experience from Sokoto

Yusuf Abdulazeez and Sundramoorthy Pathmanathan

INTRODUCTION

Informal street markets are influenced by diverse human agents and social forces across the globe and continue to shape millions of lives, especially in Nigeria, the most populous nation and largest economy in Africa as at 2013. In 2005, 57.7 per cent of its population of 139.8 million were self-employed and worked in the informal sector (World Bank, 2012). In 2012, 50.3 per cent of Nigeria's 166.6 million people lived in urban centres (United Nations Development Programme, 2013). Up until now, little scholarly attention has been paid to Nigeria's urban domestic and foreign migrants whose livelihoods come from the city streets where they sell livestock; fresh, smoked, fried, and dried meat and fish; raw and cooked vegetables; fresh and cooked eggs; fruit; orthodox and herbal medicines; mobile phone credit cards and accessories; spare parts for electronics and electrical goods; building and construction materials; clothes, belts, and footwear; and household utensils. Domestic migrants operate as street vendors in urban public spaces across Nigeria (Neuwirth, 2011; Nwaka, 2005; Onyenchere, 2011; Usman, 2010). In the country's porous border-states, such as Sokoto, migrants are likely to come from neighbouring states (including Cameroun, Ghana, Senegal, Benin, Chad, Mali, Niger, and Togo) whose economic development ratings by the World Bank (2011) are lower than Nigeria's.

Despite the rise in internal rural-urban migration and the influx of foreign migrants from elsewhere into Nigeria, most urban areas have neither adequate lucrative jobs in the formal economy, nor adequate basic services and social safety nets for indigent migrants. The informal street market is a key employment sector, even though it offers low incomes, which are incommensurate with the expectations of workers, and with the positive social and economic impact of their work. As Chuhan-Pole and colleagues (2011) explain:

> Africa's growth has not created enough productive jobs to absorb the 7–10 million young people entering the labour force each year . . . [M]ost Africans are not unemployed—they are working in low-productivity jobs in the informal sector. While continuing to create formal-sector

jobs, African countries need to increase the productivity, and hence earnings, of these informal sector workers, many of whom are in household enterprises.

(p. 9)

Poor urban residents such as those described above find solace in street markets, yet their progress is limited by intercommunal, ethnic, religious, sectarian, and political conflicts that claim property and lives. These precarious and dangerous circumstances are fuelled by what Human Rights Watch (2012) has described as "state and local government policies that discriminate against 'non-indigenes'—people who cannot trace their ancestry to what are said to be the original inhabitants of an area" (p. 144). Nevertheless, migrants doggedly continue to shape Nigeria's economy, through informal commerce and economic coping techniques that are underreported and unrecognised by stakeholders in the formal economy. This chapter examines the socio-demographic backgrounds, jobs, and the impact on the local urban economy of migrant street market workers in the city of Sokoto. It considers their survival strategies within the context of the World Bank's (2012) emphasis on the role of jobs in micro-enterprises and household businesses (such as small-scale retail and service outlets, and the street peddling of food, clothes, home ware, and lottery tickets) in the alleviation of poverty and in the improvement of living standards, productivity, and social cohesion.

METHODOLOGY

Sokoto is the capital city of the state of the same name in the northwest border region of Nigeria. According to the projections of the United Nations Population Fund, Nigeria (2012), the population of Sokoto state in 2013 was expected to be 4,602,298. Both the state and the city occupy a strategic position as a migration transit route and hub. Ilella in the north of the state is Nigeria's border-town to Niger, and onwards to North Africa and Europe. Sokoto city is an easy route and staging post for Beninese, Malian, Togolese, and Senegalese migrants entering Nigeria by land, some of whom then move to other parts of the country. This study was conducted in purposively selected parts of Sokoto metropolis. Sixteen neighbourhoods where migrants commonly reside were chosen as data collection areas: Arkilla Phase I and II, Bello-Way, Dambuwa, Diplomat, Emir Yahaya, Federal Low-Cost, Gandu Gidan-Dare, Gidan-Igwai, Kwannawa, Mabera-Idi, Manna, Old-Airport Sokoto-Cinema, and State Low-Cost. In these areas, Nigerians nonnative to Sokoto state and non-Nigerian migrant street vendors were individually contacted and interviewed at their business points. Their interviews were recorded and observed. Largely, the demographics and experiences of urban migrants in informal street markets in Sokoto were studied using a sequential explorative mixed methods research design strategy (Creswell, 2003). Purposive sampling informed the selection of 363 respondents (97 domestic and 266 foreign) who participated in the study. Semi-structured questionnaires were administered to the informal street marketers through an interview survey. This was supported with observation of the areas and streets where migrants operated informal street-businesses. In addition, in-depth interviews were conducted with five key informants selected to be representatives of communities in the study areas: two staff members of nongovernmental organizations (NGOs); one youth leader; one labour unionist; and a Southern Nigerian Christian community leader. Quantitative responses were collated using SPSS statistical analysis, while interviews were transcribed, conceptualised, coded (open, axial, and selective), quoted, interpreted, and discussed to demonstrate patterns in the jobs, commercial strategies, and challenges for informal urban street vendors.

SOCIO-DEMOGRAPHIC INFORMATION

This subsection outlines our findings on the migrants within our sample group and takes into account the demographic and social variables of age, sex, ethno-cultural, and religious background, marital status, living arrangements, and dependents.

Table 4.1 indicates that more than two-thirds of our respondents were male. Additionally, a clear majority (84.7 per cent) of respondents working in the informal street markets of Sokoto belonged to the 38-and-under age demographic. This reflects the active participation of young adults in the city's informal street vending business. While supporting Mkandawire's (2010) conclusion that a clear majority (60 per cent) of the urban labour force in Africa survives from the informal sector, the finding also partially corroborates the Population Reference Bureau's (2012) submission that most youth (15- to 24-year-old) employees in sub-Saharan Africa are in the informal economic sector. Also, the number of female domestic migrants working in informal street markets reflected the movement of women from home to market production noted by the World Bank in 2012. This finding also underscores the relative freedom of female internal migrants to inhabit and work in public spaces, in contrast to women native to Sokoto—particularly married women—who are constrained by the region's dominant Islamic cultural values that restrict married Muslim women's participation in informal street markets.

Nigeria's domestic migrants hail from many different states and ethno-cultural groups. About half (47.9 per cent) of the 97 domestic respondents were Yoruba from other states. One quarter (24.9 per cent) of the 265 foreign migrants are Zarma/Djerma from Niger, northern Ghana, Burkina Faso, Benin, and eastern Mali. The remainder are spread across 19 ethnic groups from West Africa. A majority of domestic migrants (57.7 per cent)

Table 4.1 Distribution of Respondents by Migration Type, Sex, and Age

							Age				Total
			18–24 Yrs	25–31 Yrs	32–38 Yrs	39–45 Yrs	46–52 Yrs	53 Yrs & above	No response		
Domestic migrants	Sex	Female	Count	8	15	13	5	4	1	2	48
			% within Sex	16.7%	31.2%	27.1%	10.4%	8.3%	2.1%	4.2%	100.0%
			% within Age	53.3%	48.4%	65.0%	31.2%	80.0%	100.0%	22.2%	49.5%
		Male	Count	7	16	7	11	1	0	7	49
			% within Sex	14.3%	32.7%	14.3%	22.4%	2.0%	0.0%	14.3%	100.0%
			% within Age	46.7%	51.6%	35.0%	68.8%	20.0%	0.0%	77.8%	50.5%
	Total		Count	15	31	20	16	5	1	9	97
			% within Sex	15.5%	32.0%	20.6%	16.5%	5.2%	1.0%	9.3%	100.0%
			% within Age	100.0%	100.0%	100.0%	100.0%	100.0%	100.0%	100.0%	100.0%
Foreign migrants	Sex	Female	Count	7	23	20	8	3	0		61
			% within Sex	11.5%	37.7%	32.8%	13.1%	4.9%	0.0%		100.0%
			% within Age	12.1%	19.8%	33.3%	50.0%	27.3%	0.0%		22.9%
		Male	Count	51	93	40	8	8	5		205
			% within Sex	24.9%	45.4%	19.5%	3.9%	3.9%	2.4%		100.0%
			% within Age	87.9%	80.2%	66.7%	50.0%	72.7%	100.0%		77.1%
	Total		Count	58	116	60	16	11	5		266
			% within Sex	21.8%	43.6%	22.6%	6.0%	4.1%	1.9%		100.0%
			% within Age	100.0%	100.0%	100.0%	100.0%	100.0%	100.0%		100.0%

Table 4.1 (Continued)

Total	Sex			15	38	33	13	7	1	2	109
		Female	Count	15	38	33	13	7	1	2	109
			% within Sex	13.8%	34.9%	30.3%	11.9%	6.4%	0.9%	1.8%	100.0%
			% within Age	20.5%	25.9%	41.2%	40.6%	43.8%	16.7%	22.2%	30.0%
		Male	Count	58	109	47	19	9	5	7	254
			% within Sex	22.8%	42.9%	18.5%	7.5%	3.5%	2.0%	2.8%	100.0%
			% within Age	79.5%	74.1%	58.8%	59.4%	56.2%	83.3%	77.8%	70.0%
		Total	Count	73	147	80	32	16	6	9	363
			% within Sex	20.1%	40.5%	22.0%	8.8%	4.4%	1.7%	2.5%	100.0%
			% within Age	100.0%	100.0%	100.0%	100.0%	100.0%	100.0%	100.0%	100.0%

Source: Fieldwork, 2011.

are Christians, whilst an overwhelming majority of foreign migrants (83.8 per cent) are Muslims. This diversity of participants reflects the multicultural context of Sokoto's street markets, where, by and large, different ethnic and religious groups live and transact business side by side in a city that is predominantly Muslim. Of the total 363 domestic and foreign migrants, almost half were married; 43.8 per cent were single; and 7.7 per cent were widowed, divorced, or separated. Out of the 176 married respondents, 83.5 per cent had dependents (relatives, friends, and biological and adopted children), 13.1 per cent indicated spouses as their only dependents, whilst 3.4 per cent had no dependents. Of the total 363 respondents, 52.9 per cent had dependents. By implication, a majority of the informal street marketers dedicate time to their businesses in order to meet the basic needs of dependents, in addition to themselves. Financial and social responsibilities to dependents are likely to limit operating capital, saving capacity, working hours, and the ability to plough profits back into business. Over half (58.4 per cent) of the 363 respondents shared apartments; 52.3 per cent with friends, and 6.1 per cent with extended family. Apartment sharing reflects the weak socio-economic status of street marketers in terms of owning property or affording single-family rental accommodation.

The findings here echo research on other Nigerian urban contexts such as Ibadan City (Cohen, 1969) and Kano City (Bako, 2003a; 2003b), which demonstrated that informal street markets since before independence in 1960 have been a major source of employment for domestic and foreign migrants. For this reason, they are frequently located in migrant-centred areas. It is therefore fair to say that street-commerce in Sokoto is representative of a national urban phenomenon that shapes the spatial and social structures of cities. This reflects socio-historical trends in the small-scale entrepreneurial activities that are the major sources of livelihoods for low-income, less (formally) educated, and less skilled segments of the Nigerian population, of which migrants make up a significant proportion.

TYPES OF JOBS AND HOURS SPENT IN INFORMAL URBAN STREET MARKETS

The types of jobs undertaken by migrants in informal street markets, and the number of hours spent at those jobs, are presented in Table 4.2. This provides some insight into their contribution to Sokoto's economy in general, and to its informal sector in particular.

Table 4.2 shows that almost half (47.6 per cent) of the 350 vendors who responded to both questions on the type of job and hours spent in work spent seven to nine hours in street-business daily. Over a quarter (31.4 per cent) dedicated 10 to 12 hours daily to their work. This illustrates the degree of energy, commitment and, in some cases, enthusiasm that workers in informal street markets contribute to urban development and

Table 4.2 Distribution of Respondents by Job and Hours Worked Daily

				Hours worked daily					Total
			1–3 hrs	4–6 hrs	7–9 hrs	10–12 hrs	13–15 hrs	16–24 hrs	
Kinds of jobs	Shop assistance/ salesperson/marketing	Count	0	7	13	9	0	0	29
		% within Jobs	0.0%	24.1%	44.8%	31.0%	0.0%	0.0%	100.0%
		% within Hours	0.0%	17.5%	7.8%	8.2%	0.0%	0.0%	8.3%
	Food/tea/coffee/drinks	Count	0	4	12	22	5	1	44
		% within Jobs	0.0%	9.1%	27.3%	50.0%	11.4%	2.3%	100.0%
		% within Hours	0.0%	10.0%	7.2%	20.0%	25.0%	9.1%	12.6%
	Farm produce	Count	0	2	6	2	0	0	10
		% within jobs	0.0%	20.0%	60.0%	20.0%	0.0%	0.0%	100.0%
		% within Hours	0.0%	5.0%	3.6%	1.8%	0.0%	0.0%	2.9%
	Transport/rental services	Count	0	0	3	4	0	1	8
		% within Jobs	0.0%	0.0%	37.5%	50.0%	0.0%	12.5%	100.0%
		% within Hours	0.0%	0.0%	1.8%	3.6%	0.0%	9.1%	2.3%
	Laundry services	Count	0	3	5	6	1	2	17
		% within Jobs	0.0%	17.6%	29.4%	35.3%	5.9%	11.8%	100.0%
		% within Hours	0.0%	7.5%	3.0%	5.5%	5.0%	18.2%	4.9%
	Jerrycan water vend/cart and wheelbarrow pushing	Count	0	4	22	15	2	0	43
		% within Jobs	0.0%	9.3%	51.2%	34.9%	4.7%	0.0%	100.0%
		% within Hours	0.0%	10.0%	13.3%	13.6%	10.0%	0.0%	12.3%

(Continued)

Table 4.2 (Continued)

		Hours worked daily						Total
		1–3 hrs	4–6 hrs	7–9 hrs	10–12 hrs	13–15 hrs	16–24 hrs	
Small-scale trading/business	Count	1	3	32	24	8	2	70
	% within Jobs	1.4%	4.3%	45.7%	34.3%	11.4%	2.9%	100.0%
	% within Hours	33.3%	7.5%	19.3%	21.8%	40.0%	18.2%	20.0%
Install/repair/mechanics/technician/electrician	Count	0	5	8	5	0	0	18
	% within Jobs	0.0%	27.8%	44.4%	27.8%	0.0%	0.0%	100.0%
	% within Hours	0.0%	12.5%	4.8%	4.5%	0.0%	0.0%	5.1%
Tailor/upholstery/shoe cobble/weave/carve/knit	Count	1	4	30	16	3	0	54
	% within Jobs	1.9%	7.4%	55.6%	29.6%	5.6%	0.0%	100.0%
	% within Hours	33.3%	10.0%	18.1%	14.5%	15.0%	0.0%	15.4%
Barb/hair plait/dressing/nail cutting	Count	1	3	24	6	0	1	35
	% within Jobs	2.9%	8.6%	68.6%	17.1%	0.0%	2.9%	100.0%
	% within Hours	33.3%	7.5%	14.5%	5.5%	0.0%	9.1%	10.0%
Others (e.g. car wash)	Count	0	5	11	1	1	4	22
	% within Jobs	0.0%	22.7%	50.0%	4.5%	4.5%	18.2%	100.0%
	% within Hours	0.0%	12.5%	6.6%	0.9%	5.0%	36.4%	6.3%
Total	Count	3	40	166	110	20	11	350
	% within Jobs	0.9%	11.4%	47.4%	31.4%	5.7%	3.1%	100.0%
	% within Hours	100.0%	100.0%	100.0%	100.0%	100.0%	100.0%	100.0%

Source: Fieldwork, 2011

sociability in Sokoto. Furthermore, the range of jobs outlined in the table underscores their relevance to the social statics and dynamics of modern Sokoto. The jobs recorded here, although broader in scope, are related to the low-income, physically demanding, and socially stigmatized jobs that Adedokun (2003) attributed to many unskilled foreign migrants in Nigeria. However, the number of hours that migrants spent at work was neglected in Adedokun and in several other studies on the migration–development nexus in Nigeria in Africa and globally (Adepoju, 1974; Afolayan, 2009; Afolayan, Ikwuyatum, & Abejide, 2011; Akinyemi, Olaopa, & Olorun-timehin, 2005; de Haas, 2006). In contrast, our study draws attention to the number of hours worked by migrants as a measure of the contribution made by informal street markets to urban development at local, national, and global scales.

As mentioned previously, the findings above were complemented by in-depth interviews with five informants (for whom pseudonyms are used) on the contribution of informal street markets to urban development in Sokoto. EangoSokNg, a 34-year-old union organiser, observed that the informal sector was an important source of employment and services, and thus had economic and social impacts on the growth of the city:

> For example, those from southern Nigeria maintain the business activi-ties of the state. Most of them engaged in buying and selling of goods . . . People from the south-west are road-side mechanics . . . As for the foreign migrants, they sell water (jerry can water), belts, fashion-clothes, and shoes (in stationed and mobile shops). Some are nail cutters (itiner-ant pedicure and manicure). Again, the domestic migrants sell prepared foods (itinerant and stationed restaurants) and engage in different busi-nesses in shops, where they sell provisions. Some go to markets, where they transact business with people. Some provide commercial cyclist services. They are doing a lot of work. [. . .] These kinds of jobs are the easiest and most accessible jobs they could afford. If am to rate their contribution, actually I will rate their contribution as excellent.

This was also highlighted by DangoSokNg, a 28-year-old Southern Nige-rian Christian community leader:

> They contribute a lot . . . they paid tax to the government. Some of them engaged in self-employment and owned shops, where they sell things and government collect revenues from them. There are water vendors and truck/wheel-barrow pushers among them. They also engaged in . . . some other businesses.

Other activities, such as fetching and vending drinkable water during scarcity, grass cutting, and small-scale trading, had a valuable social dimen-sion, as AngoSokNg, the chairman of a NGO, pointed out:

When there is problem of water, they fetch and vend it, so they are contributing to people's living. When you want them to cut grasses for you, they do. Some have trade, which benefit low-income earners, who usually patronise them.

Migrants working in informal urban street markets took on manual jobs, which were often rejected as undesirable by other workers because they were dirty, dangerous, and/or low-paid. BangoSokNg, who also works for a NGO, explained that the inaccessibility of adequate skills training and start-up capital for employment in the formal sector pushes many urban migrants to this type of labour:

They left their places to invest in business here. They are very vital, as they contribute a lot towards the society. There are different types of works, which natives hardly involve themselves, such as pushing trucks, shoe-cobbling, even NGOs—because of pride. These jobs are dominated by migrants, while they [natives] usually felt shy doing it. Some migrants are mechanics, upholstery makers, producers of sachet water and commercial transporters.

Nevertheless, the kinds of job undertaken by migrants in informal urban street markets demonstrate a diversity of economic activities open to those who lack access to public–private sector employment and start-up capital for larger scale entrepreneurship.

SURVIVAL AND INFORMAL STREET MARKETING STRATEGIES

Strategies and instruments deployed in the street market to secure customers' patronage include appeals to the social (courtesy, friendship, family, kinship ties), the ethno-cultural (tribal dress, greetings, languages) and the religious (utterances about belonging to the same belief system). Other strategies noticeable in Sokoto's street markets are those that capitalise on the joint demand for related products and services either inside the same shops (tailors, fashion designers, weavers, knitters) or hired and erected street stalls close to each other (iron welders, panel-beaters, tyre experts, mechanics, repairers of electronics and electrical appliances, spare parts traders). Itinerant peddlers of raw foodstuffs and of prepared foods ensure that they are located close to each other.

Our research found that migrants who operate in informal street markets in Sokoto lived amicably in the clustered, congested, and overpopulated areas of the city where they mostly conduct their businesses. The clustering of the mixed social, ethno-cultural, and religious features that define the multicultural character of these residential areas could, for some vendors, be a marketing strategy that attracts customers. Some migrants in the informal street markets lived and conducted business in inner-city areas of Sokoto that are not migrant enclaves, which implies that mobility and the extension of commercial activities beyond migrant neighbourhoods can also be an objective. As part of the set of coping strategies for urban life, workers in informal markets are goal driven and are unashamed about the status of their jobs. AngoSokNg confirmed this:

> Most of them are not lazy . . . The strategy they adopt is courage. They also have hope, wherever you see them, even better than those who are not migrants. When migrants arrived, their minds are built on hopes, so whatsoever they can do they do them. They don't see jobs as degrading, but natives see jobs as degrading. That is why they become great and even greater than those [natives], whom they met.

This viewpoint was shared by DangoSokNg:

> They [migrant informal street marketers] cope based on their activities, the little earnings and jobs, which they are doing. So, they are coping according to their works . . . When they come to a country . . . they have objectives and goals. So, they don't feel ashamed of taking any kind of job that they laid their hands on. So, they are after achieving their objectives and goals. No matter the difficult nature of the jobs, they rarely see it, as difficult jobs, even if the citizens said they don't like the jobs, the migrants will do those jobs . . . They cope based on what they are pursuing . . . I've seen migrants that married to Nigerian citizen. They don't think of going back to their country, as they felt the host country is enriched with what they need compared to their country.

This study consolidated the findings of Cohen (1969, p. 16), that participation in informal commerce is an important economic and social coping strategy for (internal) migrants in Nigeria. Cohen (1969) also documented other coping strategies for migrants, such as the practice of local customs, mixing with the local population, and familiarising themselves with local urban politics; although, as Resnick (2011) points out, politicking is frequently a risky activity for many of Africa's urban poor, particularly in opposition strongholds. In addition, migrants in Sokoto's informal street markets joined trade, ethnic, religious, and hometown associations as a means of social survival. In this light, CangoSokNg, a youth group leader, stated:

> Migrants show feeling for each other. When they come, they identify themselves with those who are originally here before the arrival of the new migrants.

According to BangoSokNg, social capital (consultation with native-born and long-term residents; making friends; marriage to locals; building relationships with members of the same religious, cultural, and ethno-linguistic backgrounds in their communities) is a crucial means of coping with the exigencies of urban life.

> If migrant comes to a place or finds himself in an unfamiliar environ-ment, he meets the indigenes or nationals to put him through, while he relates and interacts with people within the community for first-hand information on how the society operates.

Importantly, street markets provide an opportunity to mix with and secure the confidence of locally born residents so that many of these forms of social capital can emerge. This was expounded by EangoSokNg:

> they engaged in friendship . . . They are very friendly with their host community . . . They are also living peacefully with their host. These are the most important things that made them to live happily in the community.

As we found in the residential areas of Sokoto where migrants make up the majority, this social capital added value to coping mechanisms amongst communities of informal street vendors, and consolidated multicultural practices in the city.

CONCLUSION

The experiences of migrant workers in informal street markets and their contributions to the urban economy have been underresearched by aca-demia, overlooked in the literature, and sidelined in the social safety nets established by governments and development donors, yet our findings show that migration to Sokoto has had a significant impact in structuring the economic, socio-demographic, and ethno-cultural patterns of the city. The findings in this chapter emphasise the need to address what Black and Skel-don (2009) have identified as inadequate measures and tools in the collection of data, and the weak assessment and understanding of diverse types of data (e.g. data having to do with women) in the migration–development thesis. This study has opened up a new field of research: urban migrants in informal street markets. It is right to express at this juncture that further studies on this issue need not only be encouraged, funded, and conducted, but also that urban development partners ought to take adequate measures to recognise the role of migrants in informal urban economic and social arrangements. Again, it is vital that findings on migrants' contributions and challenges be converted into policy and translated into viable plans for sustainable devel-opment in Nigeria, Africa and, indeed, globally.

REFERENCES

Adedokun, O. A. (2003). *The rights of migrant workers and members of their families: Nigeria.* UNESCO Series of Country Reports on the Ratification of the UN Convention on Migrants. Retrieved from http://unesdoc.unesco.org/images/0013/001395/139534e.pdf

Adepoju, A. (1974). Migration and socio-economic links between urban migrants and their home communities in Nigeria. *Africa: Journal of the International African Institute, 44*(4), 383–396.

Afolayan, A. (2009). *Migration in Nigeria: A country profile 2009.* Geneva: IOM.

Afolayan, A. A., Ikwuyatum, G. O., & Abejide, O. (2011). *Dynamics of internal and international mobility of traders in Nigeria.* Report for the MacArthur-funded project "African Perspectives on Human Mobility". Retrieved from www.imi.ox.ac.uk/pdfs/research-projects-pdfs/aphm-pdfs/nigeria-2011-report

Akinyemi, A. I., Olaopa, O., & Oloruntimehin, O. (2005). *Migration dynamics and changing rural-urban linkages in Nigeria.* Retrieved from http://iussp2005.princeton .edu/download.aspx?submissionId=50208

Bako, A. (2003a). Opportunism or business strategy: Perspectives on Igbo commercial dynamism in Kano in the post civil war years. *Humanities Review Journal, 3*(1), 58–66.

Bako, A. (2003b). The governmental roles of religious organizations among 20th century Yoruba migrant communities in Kano. *Journal of Cultural Studies, 5*(2), 332–342.

Black, R., & Skeldon, R. (2009). Strengthening data and research tools on migration and development. *International Migration, 47*(5), 3–22.

Chuhan-Pole, P., Christiaensen, L., Angwafo, M., Buitano, M., Dennis, A., Korman, V., Sanoh, A., and Ye, X. (2011). *Africa's pulse, 4.* The World Bank/Office of the Chief Economist for the Africa Region.

Cohen, A. (1969). *Custom and politics in urban Africa: A study of Hausa migrants in Yoruba towns.* London: Routledge & Kegan Paul.

Creswell, J. W. (2003). *Research design: Qualitative, quantitative and mixed methods approaches* (2nd ed.). Thousand Oaks, CA: Sage.

de Haas, H. (2006). *International migration and national development: Viewpoints and policy initiatives in countries of origin: The case of Nigeria.* Migration and Development Series Report 6. Retrieved from www.heindehaas.com/Publications/Hein%20de%20Haas%202006%20Nigeria%20migration%20and%20development.pdf

Human Rights Watch. (2012). *World reports 2012: Events of 2011.* New York: Seven Stories Press.

Mkandawire, T. (2010). How the new poverty agenda neglected social and employment policies in Africa. *Journal of Human Development and Capabilities, 11*(1), 37–55.

Neuwirth, R. (2011, September). Street markets and shantytowns forge the world's urban future. *Scientific American.* Retrieved from www.scientificamerican.com/article/global-bazaar/

Nwaka, G. I. (2005). The urban informal sector in Nigeria: Towards economic development, environmental health and social harmony. *Global Urban Development, 1*(1), 1–19.

Onyenchere, E. C. (2011). The informal sector and the environment in Nigerian towns: What we know and what we still need to know. *Research Journal of Environmental and Earth Sciences, 3*(1), 61–69.

Population Reference Bureau. (2012). *Adolescents and young people in sub-Saharan Africa: Opportunities and challenges.* Washington, DC: Population Reference Bureau Publication.

Resnick, D. (2011). In the shadow of the city: Africa's urban poor in opposition strongholds. *Journal of Modern African Studies, 49*(1), 141–166.

United Nations Development Programme. (2013). *Human development report 2013: The rise of the south: Human progress in a diverse world.* New York: UNDP.

United Nations Population Fund, Nigeria. (2012). UNFPA in Sokoto state: Population projections. Retrieved from http://nigeria.unfpa.org/sokoto.html

Usman, L. M. (2010). Street hawking and socio-economic dynamics of nomadic girls of Northern Nigeria. *International Journal of Social Economics, 37*(9), 717–734.

World Bank. (2011). *The little data book on Africa 2011.* Washington, DC: World Bank.

World Bank. (2012). *World development report 2013: Jobs.* Washington, DC: World Bank.

5   Sounds of the Markets

Portuguese *Cigano*[1] Vendors in Open-Air Markets in the Lisbon Metropolitan Area

Micol Brazzabeni

Just look at the prices on the labels, yes ladies, six euros, look at the price that were on the labels, my ladies, this is not a joke, it's national, it is not Chink. It is nationaaal and it's only six euros, come on . . .

Perfume shop! Come, come and see, come and smell, don't pay anything. Smell it well, smell it well . . .

Eeenjoy, today these *[things]* are giifts, they are for the poor, the rich are already well pleeeased . . .

We are in criiisis, we are in criiisis, we only have criiisis in Portugaaall . . .

Passos Coelho is doing that *[Passos Coelho is the Prime Minister]*. . .

And what is the story of the *cigano* women who managed to steal it? This was stolen this morning . . .

100% cotton, look, pssssss, come to see, one and a half euros, this comes from the North *[Portugal]*, look, look, the label does not mislead, one and a half euros, 100% cotton, one and a half euros, who does not take this awaaay? All at one and a haaalf euros, giiirlss, psssss, chic garments . . .

Touch and see, long sleeved, three euros. Hey, feel it, it's a retail price, with sleeves, it's not seven and half, not ten, nor five, it's threee eurooos . . .

Look, the heat is coming, you want these garments, and these are the last ones, because the *cigano* woman is going to Brazil now . . .

Look, don't put it in your bag without paying, the *cigano* curse will fall on you, a week without going to the bathroom . . .

Look girl, rummage, the honey is underneath, you have to stir everything up . . .

Look, this is the Pingo Doooce promotion: 'Pingo Doce, come in, 50% discount' . . .[2]

Princess, ooo, six euros, help me pay the hire-purchase of my little Mercedes . . .

Two little euros, pretty girl, this is nationaaalll, what's nationaaal is good *[this was the advertising slogan of a renowned Portuguese brand of pasta, the* Nacional *pasta]*, make the most of this—it's a brand product.

The shouts and calls transcribed above are just a sample of what I recorded during my fieldwork in open-air markets in the Lisbon metropolitan area. Noise and hubbub, in other words, the shouts that characterize Portuguese *cigano* commercial practices in open-air markets are the first things noticed by people passing by: customers, political and technical local agents, and anthropologists. Shouts and calls are a constitutive and fundamental aspect of the economic activities of street selling; indeed, almost all street markets are noisy (as well as colourful, sensual, and smelly, among other things). Sometimes, newspaper articles tell us about the application of special laws to stop shouts in street markets. This happened in 2008 in the old market of Hexham town in England[3] and in 2012 in Istanbul city.[4] However, as a merchant has said, "you're not a market trader if you don't shout," and a passing customer confirmed that "there's no joy in a market shrouded in silence."[5]

My first decision when I began my fieldwork in the four open-air markets was to lose myself in the markets' soundscapes and take these sounds as my ethnographic starting point. I chose to focus on and record the shouts and calls of *cigano* vendors, especially those that might not only say something about their commercial communicative style, but also about their 'being-in-the-world.' Both from the theoretical and methodological point of view, the anthropology of senses provides the appropriate analytical tools to analyze the sensorial density of street markets, especially when it refers to the selling activity of *cigano* vendors. I would stress that the idea is not to observe the Other through the sensorial lens but to understand how others present, attribute meanings, and live their 'sensory order' as a social construction.[6]

I am not attempting to provide an analytical framework in which *cigano* vendors would represent an ethnic niche of the open-air markets. In fact, their commercial styles could be observed not just as a distinctive manner of bringing dynamism and managing the competition between each other, but mainly as performances of compressed images, stereotypes, and representations through which they place themselves in the world. In this specific context, they produce a kind of "moral economy" (Browne, 2009) about the local economic panorama, in particular the economic and social crisis.

Roger Abrahams (1981) defines open-air markets as "display events" par excellence, considering them "public occasions [. . .] in which actions and objects are invested with meaning and values are put 'on display' " (cited in Bauman, 2004, p. 58). Richard Bauman (2004; 2008) writes about the existence of a performative economy in open-air markets, represented by shouts and calls, the oral advertisements of the sellers, negotiations, and gestures. He suggests understanding markets as modern cultural performances, and as ritual events that are intensely marked by symbolic expression and language. In this sense, we can use shouts and calls as analytical objects, both theoretically and methodologically.

In my work, I assume that "artful communication," to use Bauman's (2008) description of the specific poetic performative competence that produces a certain performative act, can be recognized intuitively. *Cigano* vendors almost reduce the other vendors to silence; it seems as if the communicative skills of *cigano* sellers have somehow been trained, learned, shared within their family groups; they seem to be competitive, or, at least, dialogic. But unlike Bauman, my interest does not lie in the formal organization or characteristics of shouts and calls as poetic expression and the practical effects they produce. Neither do I analyze the shouts and calls merely as 'texts' to be decoded; indeed, they would be extremely ambiguous and paradoxical as texts. Instead, I follow the Austinian "speech acts" approach which is the perspective that words or statements perform a specific action. They make things happen, "bringing something into 'being' " (Butler, 2010, p. 151), or performatively allow certain things to happen within a given context and in certain circumstances (even instances of luck), producing effects in the world, and "altering an ongoing situation" (ibid.).

Thus, shouts and calls are performative. They do not represent the world, but act upon, and in, the world. They shape the course of the events, introduce, state, and share meanings in a specific context and with the public they attract (the public is not just a mere consumer, but a 'Black,' a Portuguese, a Cape Verdean, a friend, another *cigano* vendor or customer).

The reflection presented here is the preliminary result of my ongoing fieldwork in four weekly open-air markets in the Lisbon Metropolitan area: *Feira de Carcavelos*,[7] every Thursday; *Feira das Mercês* and *Feira de Monte Abraão*,[8] every Saturday; and, finally, *Feira do Relógio*,[9] every Sunday. Each one has different characteristics, especially in terms of location, size, renown, and customers. Moreover, the rates set by municipality authorities in each market for occupancy are subject to national political and economic change or choice.[10] However, the goods sold in these markets are a common factor, as are the noteworthy majority presence of Portuguese *cigano* dealers. It could be said that Portuguese *cigano* vendors represent between 70 per cent and 90 per cent of all the traders in these open-air markets.[11] The majority sell clothing, shoes, underclothing and socks, cheap jewellery, perfumes, clocks, cosmetics and body products, bags, and counterfeit CDs or DVDs. Some sell fabric, curtains, swimwear, and towels; occasionally, some sell books, small or decorative ware, and antiques. *Cigano* vendors do not sell food.

MULTI-SENSORALITY IN A DAY AT THE OPEN-AIR MARKET

Whilst men stand in front of or behind their stalls, shouting, selling, and managing commercial relations, women traders often sit in a small chair or stand on top of the stalls while they call the attention of the clients, shouting and clapping their hands in a remarkable and emphatic way. Alternatively, they walk backward and forward behind their stalls, never losing sight of the merchandise or the customers.

There are different aesthetic styles of organising the stalls that depend not only on the individual taste and personality of each vendor, but also on the type of product for sale. Cheap brands or second hand clothes are piled up on the stall in a disorderly and amassed manner. The traders rummage the merchandise and sometimes throw it up in the air, inviting people not to be afraid to do the same in order to find what they are looking for.

Cigano vendors create an acoustic place through their shouts. The aim is to appeal to people to discover directly what products are really worth, to involve themselves "in terms of visual, auditory, olfactory, tactile, spatial, cognitive, and behavioural engagement with the goods and social engagement with the vendors in a relationship of exchange" (Bauman, 2004, p. 66). Shouts and calls can be analyzed as multisensory performances. Shouting loudly and contemporaneously provides the trigger that activates the 'sensual' commitment between people and commodities in the exchange relationship. The 'look but don't touch' refrain, evoked by Susan Buck-Morss when talking about the figure of the *flâneur* (cited in Howes, 2005b, p. 287), here becomes 'don't just look at! Touch, rummage, feel' and, when relevant, 'smell.' Senses are the medium through which we experience the marketplace as space, as commodity, as social and economic relationships, as having values, moralities, and social significance.

Thus *cigano* vendors appeal in their economic activity, not only to the auditory and visual senses but also to the tactile and olfactory senses. The latter are historically and anthropologically associated with primitive cultures, the lower classes, the subordinate groups or minorities, and finally the Other (see Howes, 2003; 2005a). Not by chance, *cigano* vendors invoke the ignoble and uncivilized olfactory sense to insult Chinese products and the consumers of these products, advising them that buying and using them is 'shameful'—China, Chinese people, and Chinese products are considered the first economic rival and the 'enemy.' Thus, Chinese commodities "smell bad," they "smell like chicken," and "you can tell just by touching."

Howes (2005b) analyses the transition to the "sensual logic of late capitalism" through the paradigm of the "hyperesthesia": that is, the sense of 'emplacement' that people feel is created by the role played by the interconnected senses, and which depends on "sensory values produced and promoted by consumer capitalism" (p. 8). This perspective is thought-provoking for situating and understanding the agency and the power of *cigano* vendors, when creating what Haug (1988, cited in Howes, 2005b, p. 288) defines as "multisensory marketing," whose aim is to stimulate product differentiation. In this sense, the emplacement of *cigano* dealers in the marketplace is the specific agency. Unlike other contexts of the social, political, and economic life where they are withdrawn, they enter the market as if this economic sector is a privileged and legitimate space where they can take part and make a statement.

When we talk about agency and power materialized in the act of selling, we are close to the meanings and feelings Hungarian Gypsies give to their experience in the horse trade with the peasants. According to Stewart (1992), "for Gypsy men the horse-market retains its attraction since it is in their trade that Gypsy men achieve their most satisfying victories over the 'arrogant' (*barimasko*) peasants" (p. 109). Portuguese *cigano* vendors consider themselves somehow to be 'morally superior,' as they mediate between the market and people-customer,[12] and are "good at what we do," managing the whole business process: looking for, buying, handling, selling, risking, gaining, cheating, and taking advantage. It is very likely they are expressing here what they think is the "*cigano* way of doing," their ways of operating in the marketplace, and particularly their relationships with the customers and the boundaries they draw between *ciganos* and non-*ciganos*.

It is exciting to think that *cigano* sellers can in some way be social and economic protagonists, or be part of the global consumer society of late capitalism, especially if we consider two aspects. First, as Cross (2000) notes, street vending is traditionally associated with the "premodern, traditional economic order that survives only on the fringes of modern society" (p. 30). In Portugal, for instance, experts on social exclusion argue that street markets will necessarily disappear. This makes state intervention urgent, in order to retrain the people involved, namely Portuguese *cigano* families, for other economic activities, and thus to reconvert them to more appropriate jobs. However, as we can observe, there is a clear global trend toward the flourishing of informal economies and "informal formality" (Cunha, 2006). Second, the economic flow where Portuguese *cigano* traders are involved is not just a local and marginal circuit. On the contrary, as I have briefly shown, it is clearly connected with the broader economic global context. For that reason, it seems to me that *cigano* traders no longer, or not only, fill imperfect and irregular markets, as Okely claims in her work (1979). As Applbaum (2005) suggests, "open-air marketplaces can come to represent a peripheral niche [. . .] within the larger market system" (p. 277) as "*concurrent* economic processes" (Mandel, cited in Applbaum, p. 277, original emphasis).

WHO SELLS WHAT: "THE SOCIAL LIFE OF THINGS"

Although people I worked with are similar in all the markets,[13] they differ greatly in the type of goods they sell: a *cigano* widow by herself who sells "Chinese" trousers especially for women; a *cigano* woman who sells "Indian" clothing by herself or with her husband; and finally, a couple who, along with their married sons, sell "Portuguese" brands of clothing.

Selling different kinds of commodity is not merely a commercial choice. It also means choosing different forms of supply, a very different attitude in the market—including different kinds of shouts and calls. These reveal socio-economic and commercial differences between *cigano* vendors in term of prestige—the ability to do certain business, and the predisposition to manage risk—and in terms of the family's wealth and, thus, the availability of money.[14] Supply can be categorized in this way.[15] A lot of *cigano* traders get supplies from the Chinese, Pakistani, or Bangladeshi wholesale warehouses in Lisbon, buying a few pieces of various designs and sizes because each piece is expensive. Some buy and sell counterfeit stuff—they may buy directly from *cigano* traders or not, or from the few factories that still produce counterfeit goods, especially in the North of Portugal. Or they can manufacture them with the help of needlewomen. Some buy and sell branded, or otherwise, second hand clothes and shoes. Some, with more wealth and recognized business expertise, sell clothes brands that they buy in large quantities from the Northern textile factories or from the more expensive local Portuguese wholesale warehouses.[16] Just to give an example, the couple involved in the brand clothes business can buy each piece in the factories for 2 euros and sell them for 3.5 euros in the market. The woman who sells Chinese trousers can buy a pack of at most six pieces of different sizes, paying 6 or 7 euros for each piece. She sells them in the market for 10 or 12.5 euros. I suggest observing the practice of selling as a complex image, a circuit constituted by people, objects and contexts: the merchants, the shouts and calls, the commodities, the customers, the marketplaces, the narratives, or what I call rhetoric.

Appadurai (1986) is correct to assert that goods not only have a specific "biography," but also a "social life," which assigns meanings and values to people, processes, and actions. In the exchange relationship, shouts and calls, and commodities are linked and circulate like objects. They become meaningful to each other because they gain existence together in this specific context, as they are mutually constitutive and their significance does not preexist their social life. As Appadurai claims (1986), it is a specific "economic exchange [that] creates value" (p. 3).

What I observed is that things happen recurrently in the production of this kind of ephemeral event. The national product, be it Portuguese or not, is emphasized; the origin is hidden or invented so as not to say that it comes from China—paradoxically even when the actual provenance of the goods contradicts the slogan. The Chinese product is openly defamed. Finally, the material is praised; for example the cotton of Indian clothes is praised even though it is not a national product, but because Indian cotton is good.

In this framework, my perspective is that *cigano* shouts and calls activate the social life of the commodities and establish not just an economic 'reality,' but also a significant articulation between people and things (Miller, 1998). Thereby, as shouts and calls combine traders with their products, they manage different "regimes of value" (Appadurai, 1986) inside the marketplace among buyers and sellers, and among *cigano* traders themselves. We can say that shouts and calls have a "practical mediatory role" (Gell, as quoted by Rose, 2007, p. 217) that becomes part of the commodity itself, and "professional marketers of course are aware of this connection and effort to incorporate this into their strategies" (Applbaum, 2005, p. 276). They are like triggers that unleash attributions of meaning and social and economic process as well as relationships.

PERFORMING AND SELLING MORALITIES

Since the 2008 onset of the current financial crisis, scattered rhetoric, political discourses,[17] commercial events,[18] and Internet sites[19] promoting economic protectionism and cultural 'patriotism' aim to protect and encourage the national economy, enhance the national feeling of identity belonging,[20] and trigger the so-called phenomenon of 'beggar-thy-neighbour policies.'

Cigano dealers are themselves consumers of these narratives and ideologies, and "they are simultaneously their contingent producers" (Fox & Miller-Idriss, 2008, p. 546), selling them in their economic activities. They produce a kind of economic and moral compendium of what it means to be Portuguese, and of what Portuguese people need to do to save the local economy—what I call the rhetoric of the national product. What *cigano* merchants do in their commercial practices is to constantly update shouts and calls with everyday Portuguese social, economic, political, and commercial events. Thus, they are entering and participating directly as actors in the contemporary flow of socio-economic life—shaping it,[21] thanks in particular to the quality of immediacy, and consequently the flexibility and adaptability, that shouts and calls hold.

In this way, the relationships *cigano* sellers establish with their clients is based on an emotive intersubjectivity within a frame of economic and social responsibility, finally leaving the choice to the consumers' agency. Prices are low; they are priced for poor people, but at the same time they are good quality products. They are store goods, and you can really find good commodities if "you are able to search and recognize the quality." Ferrari's (2010) analysis about fortune telling among the *calins* (the *cigano* women) in São Paulo points precisely in this direction: The relation works well and gains power if it "is impregnated with an emotional charge that flows from the *cigano* woman to the *gadje* (the non-*ciganos*) and vice-versa," and able to affect the other (p. 192, own translation).

The uncertainty of bazaar exchanges seems to be a fundamental characteristic, and is why bazaars are so emotionally and sensually charged. According to Geertz (1978), "in the bazaar information is poor, scarce, maldistributed, inefficiently communicated, and intensely valued" (p. 29), and people try to orient themselves on product quality, prices, and new fashions. Besides the 'intrinsic' confusion we can generally ascribe to marketplace contexts, the uncertainty in the present case may stem particularly from two interlinked elements. The first is the ethnic identification of the seller as *cigano*; this identification is never neutral. The second is the ambiguity about the characteristics of the products and the 'authenticity' of the message promoted by shouts and calls. In fact, shouts and calls do not resolve these dilemmas—and it is also important to stress that they do not say everything about the processes involved in the market exchanges. Their main role is to advertise goods and draw customers' attention.

What happens is that shouts and calls work at the borders of the exchange relationships between *cigano* dealers and non-*cigano* customers, amplifying the aspects emphasized above as the main focus of uncertainty. Making use of national ideologies, applied to the products in a different way and with different strategies, and "trading" common "stereotypes" (Okely, 1979) produced by non-*cigano*, *cigano* vendors stimulate an 'ethnicization' of the objects, and therefore of value.[22] The commodification of negative ethnic stereotypes functions here as a positive brand, so that we are faced with the paradox that *cigano* traders promote themselves as supporting the national identity.

ACKNOWLEDGEMENTS

I would like to thank Peter Berta, Valerio Simoni, and Luís Vasconcelos for their insightful comments on a draft version of this manuscript; and also the CRIA seminars that are a fruitful occasion to share scientific works. I am also grateful to all the people who were available during my fieldwork and who offered me their time and friendship. Thanks are also due to the Foundation for Science and Technology (FCT, Lisbon) for the fellowship support to my postdoctoral research and to the Centre for Research in Anthropology (CRIA-IUL), to which I belong. All shortcomings remain, of course, my own responsibility.

NOTES

1. Above all, it is important to clarify that I decided to adopt in this article the Portuguese word *cigano/a*" and, in its plural form, "*ciganos/as*" when it is employed as a noun, and the word "*cigano*" when it is employed as an adjective. This is the term used by Portuguese *ciganos* (Gypsies) to name themselves, as well by non-*cigano* people.
2. This is the promotional refrain of *Pingo Doce*, the largest supermarket operator in Portugal. On May 1, 2012, the owner decided to do a big promotion offering a 50% discount on everything for purchases of at least 100 euros. Portuguese people—including *ciganos* from what I heard them saying the week after—participated massively in this mega-promotion, triggering great controversies in the following days and months.
3. See www.dailymail.co.uk/news/article-1046220/Stop-shouting-Market-traders-told-noisy-selling-giving-office-workers-headache.html
4. www.todayszaman.com/newsDetail_getNewsById.action?newsId=272311
5. www.guardian.co.uk/world/2012/feb/13/istanbul-market-traders-shouting-ban

6. The Portuguese word for these shouts and calls is 'pregões' and the action is 'apregoar,' that is, to announce, proclaim, publicize, or even, boast. However, Portuguese *cigano* vendors use the word 'gritos,' that is, literally 'shouts.'

7. In addition to a local reputation, it was also known in Lisbon and among tourists as a trendy market with quality and counterfeit products.

8. *Feira das Mercês* is a local open-air market that mainly supplies people living in its close proximity, largely Brazilians, 'Black' people, and migrants from East Europe (namely, from Angola, Guinea-Bissau, and Cape Verde). *Feira de Monte Abraão* is its direct competing market which is frequented weekly not only by residents but also by customers from Lisbon or surrounding neighborhoods. The cacophonic sound of the fair is easy to hear as soon as you leave the train station. Similar to *Feira das Mercês*, customers are mainly Brazilian and 'Black' people, migrants from East Europe, and Portuguese of low economic status. According to the last census in 2001, the percentage of residents from Angola, Cape Verde, Guinea-Bissau, Brazil, and more recently, East of Europe, represents 6.5% of the entire population in the municipality of Sintra (part of the above mentioned Lisbon metropolitan area), which includes the two municipalities of Monte Abraão and Mercês.

9. It is the most popular open-air market in Lisbon city and the biggest for food, clothing, and daily commodities. It is located on the outskirts of the city, namely on a highway closed to traffic during the event. It has no fencing, so people can access from all sides: top, down, laterally, in the middle, so that there is constant and confused movement. It extends for 4 km, divided in two parallel avenues, with around 700 fixed stalls and roughly 100–150 occasional ones (*cigano* vendors have at least 70% of the total).

10. In most cases, the charges are set on the basis of the space that stalls occupy per square foot; in some cases, they are fixed prices but this is not a fair policy. These rates are the State revenues for the upkeep of the markets; but they are also the greatest disincentive to work in the markets: a family which sells in three weekly open-air markets could pay up to 700 euros a month in fees, depending on the markets.

11. These data are a rough estimate given by presidents of municipalities and municipal technicians I interviewed; there are no specific statistical data on how many *cigano* vendors work in each open-air market.

12. Some of them say "we are like Robin Hood: we steal from the rich to give to the poor."

13. Except *Feira de Carcavelos* where I have only recorded sounds so far.

14. It is interesting to note that in almost all cases the value behind and associated to the business is not to have money to make and accumulate more money, but to keep the trade running, to 'untie the money,' similar to the concept of "making the money circulate" that Stewart (1994) finds among the Hungarian Gypsy horse traders.

15. This issue will be object of further investigation because it is a sensitive one in the whole commercial process; it involves a more intimate relationship with people to understand the economic and relational circuits that each family sets, which they often keep 'secret' for competitive reasons or to be embedded in what we can call an informal or illegal circuit.

16. A deeper analysis of the phenomenon is necessary, but suffice to say here that the Portuguese textile production and market were radically changed and, in the opinion of many vendors, prejudiced if not ruined by Portugal's entry into the European Community, and by the liberalisation of products from China, and by the increase of wholesale warehouses and retail shops (Chinese traders do not work directly in fairs or open-air markets).

17. See Portuguese President Cavaco Silva's discourse in 2011 (www.jn.pt/paginainicial/interior.aspx?content_id=1847481&page=-1) or Prime Minister Passos Coelho's discourse in the 'Made in Portugal' conference in 2012 (www.portugal.gov.pt/media/413644/20120112_pm_internacionalizacao.pdf).
18. See for example, the sequence of the so-called *Mega Pic-Nic* events promoted since 2009 by the hypermarkets *Continente* or *Modelo* where 'Portugueseness' is celebrated through a big display of local food products, local growers, Portuguese pop music, and folklore (www.posicionamentoweb.com/blog/geral/programa-mega-pic-nic-modelo/). We can compare these events to the fascist *Exhibition of the Portuguese World* in 1940 or to the contemporary *Expo 98*.
19. See, for example, these sites about the promotion of national products and their consumption: http://reinvestir-portugal.blogspot.pt/; http://www.compronosso.pt/; http://560.adamastor.org/, including Facebook pages; http://von.aeportugal.pt/. About one year ago, social movements of *Indignados* in Portugal also circulated an email about buying national products or spending our holidays in Portugal.
20. See, for instance, Leong Wai-Teng (2001) and Wallis (1994).
21. See Gudeman's (2009) perspective about "thinking [economy] rhetorically" to understand people's experience of material life.
22. See Berta (2007).

REFERENCES

Appadurai, A. (1986). Introduction: Commodities and the politics of value. In A. Appadurai (Ed.), *The social life of things: Commodities in cultural perspective* (pp. 3–63). Cambridge: Cambridge University Press.

Applbaum, K. (2005). The anthropology of markets. In J. C. Carrier (Ed.), *A handbook of economic anthropology* (pp. 275–289). London: Edward Elgar.

Bauman, R. (2004). "What shall we give you?": Calibrations of genre in a Mexican market. In R. Bauman (Ed.), *A world of others' words. Cross-cultural perspectives on intertextuality* (pp. 58–81). Malden, MA: Blackwell.

Bauman, R. (2008). A poética do mercado público: Gritos de vendedores no México e em Cuba. *Antropologia em Primeira Mão, 103*, 1–24.

Berta, P. (2007). Ethnicisation of value—the value of ethnicity. The prestige-item economy as a performance of ethnic identity among the Gabors of Transylvania (Romania). *Romani Studies, 5*(17), 31–65.

Browne, K.E. (2009). Economics and morality: Introduction. In K. E. Browne & B. L. Milgram (Eds.), *Economics and morality: Anthropological approaches* (pp. 1–42). Lanham: Altamira Press.

Butler, J. (2010). Performative agency. *Journal of Cultural Economy, 3*(2), 147–161.

Cross, J. (2000). Street vendors, and postmodernity: Conflict and compromise in the global economy. *International Journal of Sociology and Social Policy, 20*(1), 29–51.

Cunha, M.I. (2006). Formalidade e informalidade: questões e perspectivas. *Etnográfica, 10*(2), 219–231.

Ferrari, F., (2010). *O mundo passa. Uma etnografia dos Calon e suas relações com os brasileiros*. Unpublished doctoral dissertation, Faculdade de Filosofia, Letras e Ciências Humanas, Dep. Antropologia Social, Universidade de São Paulo, São Paulo.

Fox, J. E., & Miller-Idriss, C. (2008). Everyday nationhood. *Ethnicities, 8*(4), 536–576.

Geertz, C. (1978). The bazaar economy: Information and search in peasant marketing. *American Economic Review, 68*(2), 28–32.

Gudeman, S. (2009). Introduction. In S. Gudeman (Ed.), *Economic Persuasion* (pp. 1–14). New York: Berghahn Books.

Howes, D. (2003). *Sensual relations. Engaging the senses in culture & social theory.* Ann Arbor: University of Michigan Press.

Howes, D. (2005a). Introduction. In D. Howes (Ed.), *Empire of the senses* (pp. 1–17). Oxford: Berg.

Howes, D. (2005b). Hyperesthesia, or, the sensual logic of late capitalism. In D. Howes (Ed.), *Empire of the senses* (pp. 281–303). Oxford: Berg.

Leong Wai-Teng, L. (2001). Consuming the nation: National day parades in Singapore. *New Zealand Journal of Asian Studies, 3*(2), 5–16.

Miller, D. (1998). Why some things matter. In D. Miller (Ed.), *Material culture* (pp. 3–24). London: UCL Press.

Okely, J. (1979). Trading stereotypes: The case of English Gypsies. In S. Wallman (Ed.), *Ethnicity at work* (pp. 17–34). London: Macmillan.

Rose, G. (2007). *Visual methodologies* (2nd ed.). London: Sage.

Stewart, M. (1992). Gypsies at the horse-fair. A non-market model of trade. In R. Dilley (Ed.), *Contesting markets: Ideology, discourse and practice* (pp. 97–114). Edinburgh: Edinburgh University Press.

Stewart, M. (1994). La passion de l'argent. Les ambiguities de la circulation monétaire chez les Tsiganes hongrois. *Terrain, 23*, 45–62.

Wallis, B. (1994). Selling nations: International exhibitions and cultural diplomacy. In D. J. Sherman & I. Rogoff (Eds.), *Museum culture. histories, discourses, practices* (pp. 265–282). London: Routledge.

6 Subcultural Citizenship in El Chopo, Mexico City

Tony Mitchell

Situated on the Calle Camelia, between Calles Sol and Luna and the Buenavista and Guerrero metro stations, and alongside what used to be a railway yard, is the *tianguis cultural del Chopo* (*El Chopo* cultural market). 'El Chopo,' as it is known colloquially, consists of up to 150 stalls and regularly attracts crowds of around 7,000 people. Until recently it went unmentioned in the major English language tourist guides to Mexico, and the English language entry on Wikipedia consists of three brief paragraphs. A Wikipedia excursion to 'Mexican Rock Music' describes the market's origins under the heading "Melting Pot: Chopo Bazaar." In 1980 a record swap meet was set up inside the Universidad Nacional Autónoma de México's Museo del Chopo in the wake of a punk-related exhibition at the museum. Owing to overwhelming demand, the swap meet was forced to move outside into the street. Held weekly on Saturdays, the event began to attract wider attention, along with antagonism from local residents and authorities, and it "wandered in exile, facing constant police harassment" before establishing itself "semi-institutionally" (Zolov, 1999) as a site for 'outcasts' from 'official' society. According to the Wikipedia entry,

> between 1982 and 1989 the 'Chopo' (as it was now known) changed locations no less than six times, from parks to parking lots to faculty gardens, always because of pressure from officials. . . . Finally, since 1990 it has been taking place on a street behind the Buenavista Train Station, not three blocks away from the original 'Museo del Chopo' location.
>
> (http://en.wikipedia.org/wiki/Mexican_rock music

From its humble origins, El Chopo became a place of belonging, and o social and cultural practices. Young people of all subcultural persuasions could exchange records, cassettes, CDs, mp3s, fanzines, and flyers for live music gigs and political rallies related to punk, heavy metal, goth, prog rock, reggae, ska, hip hop, and most other independent or underground rock music-related genres, in both English and Spanish. Among the plethora of bootlegged and pirated CDs and mp3s from the vast array of musica genres that is found at the market, is *Ska Exitos*, a compilation of Mexican, Argentinian, and other Latino ska music by prominent musicians such as Inspector, Gran Silencio, La Maldita Vecindad, Tijuana No, Los Fabulosos Cadillacs, and Mano Negra. "Intro El Chopo," the opening track of the compilation, enshrines the market as the logical location linking these musical entities. This album in its musical scope and subcultural heterogeneity functions as a metonym for El Chopo.

URBAN FLEA MARKETS AND SUBCULTURAL CITIZENSHIP

Reed Johnson, in a 2004 *Los Angeles Times* article titled "El Chopo is the place to be for outcasts," describes the market's location as

> an aging neglected part of the city, the type of neighbourhood that Mexicans call a *barrio bravo*, a term connoting hardship, pride and wild, unruly creativity.
>
> (p. 2)

El Chopo is, as one of Johnson's informants puts it, "a space of coexistence," and has affinities with the peripheral locations of many outdoor urban flea markets, including the *Marché aux Puces* in Paris, *Porta Portese* in Rome, *Mauerpark* in Berlin, *Saint-Eustache* north of Montreal, and the *Otara* market in Auckland. Kieran Keohane (2002) observes that

> the flea markets of the world's cities are thronged with urbanites; ironists trawling for kitsch, romantics searching for scraps of the authentic, predators "bargain hunting"; and where markets have been demolished they are soon replaced by sanitized simulacra . . . The moral significance of the market in the modern metropolis is that participation in it is a civilizing process: it is this that is dangerously eliminated from sanitized postmodern simulacra of markets. In the marketplace, the flea market, the second-hand shop, the price of the commodity is not fixed. In the flea market, negotiation is normative, and this is a unique relation in the wider context of the complete penetration of rationalization and the commodity form in the city (even the price of drugs from a street dealer is not subject to negotiation). Bargaining, haggling, wheeling and dealing, compromising, and associated communicative arts and skills are developed and honed in these contexts, and, as Simmel (1971b) would insist, here are the play-forms of the discourse, liberal tolerance and democratic citizenship.
>
> (p. 43)

Criminalised practices of contraband and piracy, which predominate at El Chopo, might seem to coexist uneasily with Simmel's notions of "liberal tolerance and democratic citizenship"; at least from the point of view of legal frameworks, supported by governments and some in the music industry, which outlaw bootlegging and piracy. But working from Joke Hermes's (2005) definition of cultural citizenship as "the process of bonding and community building, and reflection on that bonding, that is implied in partaking of the text-related practices of reading, consuming, celebrating, and criticizing offered in the realm of (popular) culture" (p. 10), El Chopo could be defined as a site of subcultural citizenship that transfers these text-related practices into a more informal, 'underground' context. Stevenson (2003) has suggested that "ideas of 'cultural' citizenship need to be able to define new forms of 'inclusive' public space so that 'minorities' are able to make themselves and their social struggles visible and open the possibility of dialogic engagement" (p. 333). In keeping with this, El Chopo represents an inclusive place of belonging, visibility, and dialogue, for political, social, and cultural minorities. In many cases the participants at El Chopo mobilise a blurring of the distinction between producer and consumer, in a sphere of interactivity where entertainment, leisure, and lifestyle practices become forms of vernacular expressions of subcultural citizenship. To borrow a metaphor from Hermes (2005), they constitute "the threads from which the social fabric is knit" (p. 11). These threads are arguably linked through various subcultural practices which provide what Will Straw (2005), writing of the Canadian context, has referred to as "pathways of cultural movement":

> Vibrant networks of cultural activity may leave behind few visibly successful works or cultural milestones. It is in the movement of social energies along such networks, nevertheless, that we might seek indications of cultural achievement or vitality.
>
> (p. 184)

The physical equivalent of El Chopo's "pathways of cultural movement" is the crowded alleyways in which its numerous stalls are situated, where interaction between 'social energies' deriving from a *mestizo* of numerous Mexicanised subcultures offer a vibrant and highly successful blend of subcultural practices.

INDIGENISING SUBCULTURES

El Chopo is focused almost entirely on varieties of rock music and their *accoutrements*. Almost all the merchandise at El Chopo is 'illegitimate,' in that it is pirated or associated with marginal social activities. Unlike the prevalence of junk and detritus in other flea markets, El Chopo specialises almost entirely in subcultural fetish objects. The atmosphere of fear and menace that sometimes prevails with the presence of pickpockets and other con artists, for example at Porta Portese in Rome, is entirely absent here.

Camaraderie and friendliness predominated during my visits to El Chopo in 2004 and 2005, alongside a seemingly shared tolerance and respect among various music-related subcultures. The spirit of acceptance towards me as an outsider, a foreign music fan looking for pirated Mexican music, arguably spoke of a solidarity forged in shared struggles at El Chopo against the authorities, some of which have been documented by Alan O'Connor (2000).

The name of this "vast rock flea market" (Berthier, 2002) is connected to that of its original location. It is a museum of, and a monument to, piracy, and in his 1992 essay "La hora del consumo alternativo: El tianguis del Chopo," revered Mexican cultural critic Carlos Monsiváis has famously called the market

> a temple to Mexican counterculture . . . the result of a conference and the natural tendency for exchange, the famous *potlach* of anthropology . . .

> Every Saturday the mania spreads, the seeking out of opportunities, the apprenticeship of discographies that justify one's existence. The participants in El Chopo value discographies above any other object on earth, and the myth of the party (the orgy of the senses) over the (unobtainable) advantages of success in the corporate world. In its elegized marginality, everything and nothing is an object of commerce, and everything and nothing is an object of transgression.

> Music, the scene which is recreated and loved, conspires in the open air and is represented through the history of rock, the recordings of groups who haven't even sanctioned these recordings, the pain and fury of heavy metal, punk, post-punk, thrash, hard core, anarcho-punk, trans-avant garde, prog rock . . .

> . . .

> [El Chopo attracts] a lost tribe . . . who live the laws of excess, satiating themselves with the fetishism of the earring that protects them from convention, from their hair to their belt. Embracing their acquisitions, they boast of wildly romantic ironies. Here there is no rock band old enough, no CD new enough, and the culture of something democratised in the marginality that it never quite becomes.
>
> (Monsiváis, 1995, pp. 120, 121, 123, 124)[1]

The democracy of the marginal that Monsiváis observes echoes Keohane's (2002) "democratic citizenship," where participation in the marketplace of "bargaining, haggling, wheeling and dealing" is a levelling process that seemingly reduces distinctions of class and wealth. Here there is solidarity in marginality and the shared freedom of subcultural tribes who display respect and tolerance for each other across the borders of their dress, demeanour, and musical tastes: here a rude boy in full regalia might check out a metal band, or a b-boy may converse with a punk. Although the swap mart dynamic of El Chopo has been replaced by a stallholder and vendor *modus operandi*, an atmosphere of cooperation still prevails in the devotion to cheap pirated CDs.

The singularity and specificity of place is enhanced through references to the indigenous and to Aztec origins, as invoked by Asael Grande Garcia when he imagines the market as

> a classic Aztec place of barter . . . the only place that exists . . . in the whole world where there is room for everybody, without exception, without censorship, without authority, without police or tabloid press, without laws, and of which the only master and love is Rock.
>
> (2001, n.p.)

Tianguis is an indigenous Mexican word for "open air market" (Zolov, 1999) and further tropes of the indigenous exist in the anarcho-punk patches proclaiming "we are all indigenous" (Iten, 2002) and in the heavy metal imagery that draws extensively on florid Mexican tropes of death.

According to Eric Zolov (1999), at El Chopo rock music lives in its indigenised, pristine countercultural Mexican forms as "a signifier of cosmopolitan values and a bearer of disorder and wanton individualism" (p. 11). "In Mexico," says Zolov, "rock mattered . . . both as an instigator of modern values and as a reflection of the modernisation process itself" (p. 15). There are also countercultural residues of *la onda* (the wave), a polysemic term that reaches back to 1968, when several hundred protesting students were massacred by the army in the Tlatelolco district on the outskirts of Mexico City. In his 1976 collection of essays, *Amor perdido*, Monsiváis defines *la onda* as "the first movement in modern Mexico that, from an apolitical position, rebelled against institutionalised concepts of culture and eloquently revealed the extinction of cultural hegemony" (in Martínez, 1993, p. 159). *Onda* has become slang for trip, deal, situation, as well as for the flow of Mexican countercultural history that pours into El Chopo.

Another key traditional and translocal symbol in the indigenisation of Mexican subcultures to be found in El Chopo and elsewhere is the brown-skinned Aztec-Catholic *Virgen de Guadalupe*, the patron saint of Mexico since 1737, and traditionally a symbol of suffering Mexican motherhood and the strictures of a virulent Hispanic Catholicism, which still grips 90 per cent of Mexico in a stranglehold. The Virgin of Guadalupe was first appropriated in the 1960s by the governing Institutional Revolutionary Party, which also tried to make rock and pop music part of its programme. She has also been used as a symbol in Mexican rock subcultures and as an icon of the *mestiza* indigenisation of Mexican youth. The name of Mexican rock group la Lupita, whose albums are usually visible at El Chopo, is a diminutive of Guadalupe.

The complex trajectories of *rock en español* also find full expression and identification among the merchandise of El Chopo, along with the United States and United Kingdom musical genres that have inspired them. But as Monsiváis (1995) has pointed out, the emphasis is firmly on alternative and independent forms of music, and mainstream pop and rock is notable by its absence. El Chopo is, as one of Johnson's informants puts it, "a place where everyone can show their strangeness" (Johnson, 2004). The black velvet and makeup of the *darketos*, the Mexicanised goths, identify as followers of groups like Caifanes, an indigenised Mexican group influenced by British goth band The Cure, or of Santa Sabina, a flamboyant cabaret rock group who canonise the significant countercultural figure of the mushroom priestess Maria Sabina, considered a witch and drug abuser by the established church. Even the multilingual Jaramar, who sings rich, dark, fruity, and funereal cabaret styled ethnic music from the 12th century onwards, in ancient languages from Latin to Provençal to Galaic-Portuguese to more modern languages like German, French and Zapoteca, has reputedly been appropriated by the *darketos*. Suits and braces identify the ska rude boys, who follow groups like Maldita Vecindad y los Hijos del Quinto Patio, whose name means the Damned Neighbourhood and the Sons of the 5th Tenement. Then there are the privileged rich boys Café Tacvba, who take their name from a ritzy inner-city café of the same name. Café Tacvba might be defined in *caló*, or slang, as *fresas*, strawberries, or yuppies, but one of their most famous songs is a cover version of "Chilanga Banda," an untranslatable mish-mash of Mexican slang, originally written and recorded by Juan Jaime López in the 1980s, that referred to DF's (or *Distrito Federal*, the locally used name for Mexico City) youth subcultures.

Hector Castillo Berthier is an academic and founder of the rock outreach centre *Circo Volador* ('Flying Circus'), a thriving countercultural centre for Mexico City's gangs and *chavos banda*, with its own radio station and a plethora of cultural events, which owes its origins to an academic sociological study of Mexican youth. Hector described El Chopo to me, mindful of its origins, as an alternative 'museum' that contains vast archives of pirated musical recordings, iconography, ephemera, documentation and information about independent and underground music. In "My Generation: Rock and *Banda*'s Forced Survival Opposite the Mexican State," Berthier (2002) sketches in a more specific history of El Chopo:

By the late 1970s and early 1980s, rock had begun to leave the spaces and consciousness of the periphery to initiate its slow but steady return toward the middle classes. A seminal moment in this process was the appearance of the Tianguis del Chopo, a vast rock flea market where attendees were invited to bring their "old records" to exchange them with others, in October 1980. Co-organized by Ángeles Mastretta, director and coordinator of cultural events at the Museo del Chopo, and Jorge Pantoja, of the Pro-Music and Art Organization, the tianguis "brought together collectors, producers, musicians, and fans interested in record production and collection of any sort," especially music that was considered "rare, un-catalogued, or out-of-print." The proposal was originally conceived as an attempt to "slow down the black market's grasp on collection material" (ibid), but it served, in turn, as an incentive for the work of independent labels and producers, as well as for the promotion of old-fashioned, vinyl records and for bands lacking commercial exposure. Attendance during the following Saturdays was such that the rockeros who arrived to exchange materials greatly exceeded the organizers' expectations, and demanded that the tianguis be installed every Saturday. These rock fans and connoisseurs, in coming together in El Chopo, set the trend, almost unknowingly, for the birth of what would be the most important distribution of rock at the national level.

(pp. 347–348)

All roads lead to El Chopo, and it is itself a road, or a series of roads, criss-crossing and interrelating over three blocks, as different subcultures interact in the interests of just and equitable barter and exchange. Appropriately, the Café Tacvba CD I picked up there was entitled *Quatros Caminos*, or Four Roads, which is also the name of the last metro station on line 2 of the DF's rail network. This commonality in countercultural opposition is described vividly in a trip to El Chopo in its early days by L.A. journalist Rubén Martínez (1993):

[We] dive into the marketplace. Throngs of Mexico City youth in all manner of rockero regalia surround us: chavas in leather miniskirts or torn jeans, chavos wearing Metallica T-shirts, James Dean leather jackets or Guatemalan-style indígena threads. We walk past stall after rickety stall, scraps of splintered wood and twine holding up faded blue tarpaulins, where the vendors—young punkeros or trasheros (thrash fans), leathered heavy metaleros, Peace and Love jipitecas and the working-class followers of Mexican raunch-rock heroes El Tri known as chavos banda—sell cassettes, CDs, LPs and singles, bootlegs and imports, as well as posters, steel-toed boots, skull earrings, fan mags spiked bracelets and collars, incense and feathered roach clips. Ghetto-blasters blast Holland's Pestilence, Mexico's El Tri, Argentina's Charly García, Ireland's U2.

"*Tenemos punk, tenemos heavy metal, tenemos en espanol y en inglés tenemos al Jim Morrison y El Tri!*" yells a young vendor, exactly as any one of Mexico's army of street vendors hawks rosaries or Chiclets. His is but one voice among hundreds at El Chopo, as the sprawling swap meet is known. It's a Saturday afternoon, some 10 years after this institution was born, and the vendors tell me that the crowd of about thre thousand is on the light side. "What's *chingón* is that there's no divisions here between the different rockeros," proclaims Ricardo, a high-school kid in a T-shirt emblazoned with the logo of the punk band LARD, a Visions Streetwear beret and hip-hop hi-tops. "It doesn't matter if you're hardcore or *trashero*."

(pp. 150–151)

Ricardo's eclectic mix of punk, beatnik, and hip hop suggests a transcultural bricolage. Above all, El Chopo is a place to congregate, to display oneself, to "hide in the light," where Dick Hebdige's (1988) representation finds its most vivid evocation:

Subculture forms up in the space between surveillance and the avoidance of surveillance, it translates the fact of being under scrutiny into the pleasure of being watched. It is a hiding in the light.

The "subcultural response" is neither simply affirmation or refusal, neither "commercial exploitation" nor "genuine revolt." It is neither simply resistance against the external order nor a straightforward conformity with the parent culture. It is both a declaration of independence, of otherness, or alien intent, a refusal of anonymity, of subordinate status. It is an *in*subordination. And at the same time it is also a confirmation of the fact of powerlessness, a celebration of impotence. Subcultures are both a play for attention and a refusal, once attention has been granted, to be read according to the Book.

(p. 35)

El Chopo represents both a preservation of the historical resistant aspects of subcultures and a free flow and interaction between different subcultures. For me, the latter was epitomised by the sight of two Socialists in bright red T-shirts, featuring full-face portraits of Lenin and Marx, handing out leaflets to a punk with Nazi insignia on his T-shirt.

EL CHOPO AND AN UNDERGROUND ECONOMY

A number of Mexican intellectual accounts of El Chopo emphasise its quasi-Utopian communal collectivism, often invoking historic precedents of Mexican youth countercultures. Grande Garcia (2001) states:

[T]here exists an unwritten treaty between the marketgoers and the neighbours . . . a kind of resigned tolerance of the protest . . . the neighbours have other worries, like the consumption of beer, drugs and violence in the streets. Some urban passers-by misinterpret El Chopo if they're not familiar with this type of expression by young people, and the police generally refrain from coming into the market … El Chopo is a scene bringing together social interaction acted out by young people … marked by the characteristics of sharing a different vision of the world, a different way of seeing, living and feeling reality.

(n.p.)

But El Chopo also comfortably inhabits a sprawling, chaotic, vastly over-populated city full of unofficial salespeople in markets and on the streets. Monsiváis's (2002) essay on Mexico City, "Identity Hour or, What Photos Would You Take of the Endless City?" describes

[t]he underground economy that overflows on the pavements, making popular marketplaces of the streets. At traffic lights young men and women overwhelm drivers attempting to sell Kleenex, kitchenware, toys, tricks.

(pp. 31–32)

An extreme example of 'tricks' I witnessed was a young woman who, while holding a baby, would juggle at traffic lights in a bid for small change from motorists. The underground economy of El Chopo spills over into and blurs with overground and mainstream commodification and tradition.

The taxi driver who takes us from El Chopo to the Frida Kahlo 'Blue House' museum in Coyoacán has been doing a spot of CD shopping at the market, too. He is a slightly wasted-looking blues fan with shoulder-length hair, probably in his late thirties, who lived in Texas for a while and speaks some hesitant English. He is a fan of the late great Texas blues guitarist Stevie Ray Vaughan, who is blaring away from the stereo on the taxi's dash. It is one of those green Volkswagen bug taxis with the front passenger seat removed, and lying on the floor in the front is a Nirvana CD. We immedi-ately feel at home in his cab. Much later I read a huge front page *Sydney Morning Herald* travel feature by Catherine Keenan (2005) about Mexico City, which caused me to laugh out loud.

Never catch the green Volkswagen cabs that are everywhere—they are regularly implicated in robberies and kidnappings. If you want a cab, get someone in a hotel or restaurant to phone a tourist cab or pick one up at the few *sitios* (taxi stands). They're more expensive but safe.

(p. 5)

We had been hailing these fabulously dilapidated and eviscerated green Beetles all over the city for the past week, without the slightest spot of bother.

The Coyoacán is a fair distance southwest of El Chopo, a good half hour's drive away, but Stevie Ray Vaughan provides an appropriate raw blues rhythm which cuts through the congested streets, with their eccentric roadside salespeople flogging everything from inflatable dolls to fluffy toys and pirate CDs. The fare is just under 100 *pesos*, less than my Café Tacvba CD. We give the driver a generous tip. "You good people," he says. It is just another example of the friendly, inclusive spirit that we had encountered in El Chopo. Like our taxi ride, El Chopo continues to maintain both what Berthier (2002) has described as "a contestatory, rebellious and irreverent presence" (p. 339), and a distinctive socially marginalised identity, which will hopefully never make it into the official tourist guides.

ACKNOWLEDGEMENTS

Thanks to Moses Iten for many of the references, translations and his UTS undergraduate thesis, *The Soundtrack of Counterculture in the Context of Mexico 2002*; to Hector Castillo Berthier for a great night out in a wild Irish pub in the DF with Celtic and Anglo-American music played by Latin-American and Scottish musicians; and to Lisa Leung Yuk Ming for the company and the photos.

NOTE

1. Translated from the Spanish by Moses Iten.

REFERENCES

Berthier, H. C. (2002). My generation: Rock and *banda*'s forced survival opposite the Mexican State. In E. Zolov (Ed.), *Rockin' las Americas*. Pittsburgh: Pittsburgh University Press.

Garcia, A. G. (2001). El tianguis cultural del chopo. Retrieved from www.cultura.df.mx/2001/abr/tianguis/

Hebdige, D. (1988). *Hiding in the light: On images and things*. London: Comedia.

Hermes, J. (2005). *Re-reading popular culture*. Malden, MA: Blackwell.

Iten, M. (2002). *The soundtrack of counterculture in the context of Mexico 2002*. University of Technology, Sydney, In-Country Study Undergraduate Project.

Johnson. R. (2004, August 8). "El Chopo" is place to be for outcasts. *Los Angeles Times*.

Keenan, C. (2005, January 21). Walls of fame. *Sydney Morning Herald*. Retrieved from www.smh.com.au/articles/2005/01/21/1106110930043.html

Keohane, K. (2002). The revitalization of the city and the demise of Joyce's utopian modern subject. *Theory, Culture & Society, 19*(3), 29–49.

Martínez, R. (1993). *The other side: Notes from L.A., Mexico City, and beyond*. New York: Vintage Books.

Monsiváis, C. (1995). La hora del consumo alternativo: El tianguis del Chopo. *Rituales del Caos*. Ediciones Era, Mexico.

Monsiváis, C. (2002). *Mexican postcards* (J. Kraniauskas, Ed. & Trans.). London: Verso.

O'Connor, A. (2000). El Chopo rock market: The struggle for public space in Mexico City. Retrieved from http://ontario.indymedia.ca/twiki/bin/view/Toronto/ElChopoMarket

Stevenson, N. (2003). *Cultural citizenship: Cosmopolitan questions*. Maidenhead, UK: Open University Press.

Straw, W. (2005). Pathways of cultural movement. In C. Andrew, M. Gattinger, M. S. Jeannotte, & W. Straw (Eds.), *Accounting for culture: Thinking through cultural citizenship*. Ottawa: University of Ottawa Press.

Zolov, E. (1999). *Refried Elvis: The rise of the Mexican counterculture*. Berkeley: University of California Press.

7 The Hidden Market
London's Alternative Borough Market

Daisy Tam

Borough Market is a renowned market situated in the heart of tourist London, near London Bridge. The upscale food market enjoys a popularity which, according to cab drivers, rivals even that of the famous Westminster Abbey or Buckingham Palace. Ever since food became a highly mediatized topic, knowing where to find quality produce has become a trending subject. Borough Market is a gastronomic reference for many food lovers in London because of its traditional butchers, fishmongers, greengrocers, bakers, and cheese mongers. As the source of "exceptional British and international produce" (boroughmarket.org.uk), the market attracts the attention of many celebrity chefs, whose endorsement amplifies the significance of Borough Market in the food world.

In addition to renewed public interest in taste and quality, there is also a heightened awareness of the impact of food production and distribution. Investigative journalism and advocacy movements backed by the media have highlighted concerns for the welfare of animals, the environment, and the people who work to produce our food.[1] The increased exposure has created a public that is as well versed in gastronomy as it is in the ills associated with large-scale commercial farming. Having worked in Borough Market for five years for a local organic apple farmer, I have grown accustomed to questions about the family-run farm: what varieties of apples are grown, whether they are organic or not, what methods of pest control are used, where the farm is located, and so forth.

Borough Market and other farmers markets are hailed not only as the 'haven' for quality produce, but to a large extent as the antidote to supermarkets, which have come to exemplify the model of large scale commercial farming. Markets are perceived to be better than supermarkets on the basis of believed higher quality, more diverse variety of produce offered, and the support that is given to local, ethical, and sustainable models of agriculture and commerce. The reduced distance between producers and consumers is often reinforced through the narratives of people behind the food. In comparison to the aisles of supermarkets and chain stores, the environment in which the food is sold is more lively and personal. Visitors are drawn into the atmosphere of the market, enthused by the sights, sound and smells, photographing the displays of produce, trying to capture the market experience, and even perhaps trying to recreate the scenes with which they have become familiarized through mediums such as television.

This chapter draws the attention away from the glamorous representation of the market to offer a 'behind the scenes' look through my eyes as a trader/ethnographer, a backstage view that does not necessarily feature or tie in with the grand narrative of the market. This hidden, more informal market is at the periphery, obscured, often forgotten or overlooked, denigrated—yet is vital and constitutive of the life and effective functioning of the formal dimensions of the market—and indeed could be argued to form the basic substance of the market itself. The first part of the article offers a description of the market at closing time, presenting some of the key features of the informality that I wish to explore. Following this, I provide an elaboration on the wider implications of the case study, and argue for the importance of acknowledging and recognizing this hidden informal market.

END OF MARKET DAY

The distant shouting of "half-price fish!" signals the closing of the official market day. The pulse of Borough Market quickens once again after the lull of the afternoon trade. The final hours of the market are often a flurry of activity: traders rush around dismantling displays while simultaneously trying to sell off the last of their stock; transit vans and garbage trucks jostle for space as they back through the gates while collectors in fluorescent jackets weave through the traffic of the market to gather overflowing bins. In the midst of the hustle and bustle, a final wave of customers comes in to catch the end of the day offers—buy-one-get-one-free bread, two pies for a 'fiver'—this is the time to snap up good deals. Against the din of traders shouting their final prices, another group of regulars arrive: Kenny, who sells the magazine the *Big Issue*[2] at the market, pays me a well-timed visit for a glass of free juice; another lady waits with her trolley for a quieter moment to ask if there are leftovers that she could take. While these regulars might not be considered as key actors in the formal economy, they are nevertheless part of the fabric of market life.

Saturday is the last trading day of the week. All stock that cannot be kept until the next market day will be cleared out. Some, like bakers or fishmongers, offer end-of-day sales to get rid of the leftovers while others try to keep their produce exclusive by throwing out excess. The manager of an olive stall, for example, refuses to cut prices at the end of the day for fear of encouraging customers to frequent her stall only at closing. While the manager of the olive stall may be following the official rule, employees nevertheless try to recuperate the leftovers and secretly give them out to their trader friends after work. The food that does not make it onto the workers' tables is carefully discarded at the nearby rubbish collection point, where a quiet wink or nod indicates to those in the know which of the bags contain recoverable food. Before the collectors arrive, 'freegans' or 'dumpster divers'[3] take the window of opportunity to pick and choose from the piles of perfectly good loaves of bread, bags of olives, baskets of vegetables, and other produce.

Working as a trader in the market, I enjoy a weekly supply of olives, bread, butter, and sausages from my network of friends. In return for these favours, I give them rebottled juice from my stall. These might be offered as a gesture of good will, a sign of friendship, or in exchange for other goods; it can also be used to return favours—such as being allowed to jump the queue, or an extra discount for lunch. There is an unspoken rule that entitles all workers to a discount (although the scale of reduction is not set); how much discount one gets often depends on the friendships between traders; over the years I worked in the market, the changing details of the purchases reflected the progressive strengthening of social bonds. This was also observable in the hierarchy of choice in the food chain of leftovers in the informal economy of exchange—I went from recuperating whatever was left after everyone else had concluded their transactions, to a position which allowed me to place my requests at the beginning of the day and have the first pick.

Based on these leftovers and excesses—some by-products of regular trade, some produced by market workers alongside their official activity—there exists another less visible stratum: a subterranean informal economy of exchange, using leftover goods as currency, operating amongst different groups and networks of people. To a certain extent, this informal economy of reciprocating gift exchange resembles the gift economy elaborated by Marcel Mauss (2006) in his study of the Polynesian tribal society. The important point that Mauss makes is that the offering and receiving of gifts might appear voluntary, free, and disinterested, but in fact every gift given demands an obligation to receive, and a further demand to return it—either, depending on the tribes, with gift amounts of equal or excessive value. In other words, gifts are never given "freely," "never unrequited" in that there is always an "obligation" to return gifts to the extent that "economic self-interest" is always involved in the exchange (Mauss, 2006, p.4).

In the market, there are cases of calculated pleasantries and sometimes even forced trades where unwanted goods are given in an attempt to contract the other into a system of exchange. For example, Ben, who works in the raclette stall stormed over one day complaining about a tuna supplier who tried to force some fish on him, so as to get a discount on his cheese sandwiches. The premeditated offer was imposed upon Ben, who "didn't want fish in the first place," in order to contract him into a transaction and oblige him with the need to *revanchieren* (to return/reciprocate) (Mauss, 2006, p. 8) the 'favour' with a discount. While there are certainly differences in the culture of gift-giving between Polynesian tribes and the Borough Market,[4] the important point to take from this is that all exchanges are undeniably contractual in spirit—the obligation to receive and reciprocate forces the individual to enter into a system of exchange such that "each gift is part of a system of reciprocity" (Mauss, 2006, p. xi).

TRICKS AND TREATS

These informal exchanges are, from an official perspective, illegitimate; a sub-socio-economic sphere construed by gift exchanges made on the sly. The 'tricks' and 'diversions' that survive in the interstices of the mainstream are important according to Michel de Certeau (1988), whose attention to these sleights of hand gives significance to the clandestine practices that could go unnoticed within a purely Maussian model. In between packing away stock and serving customers, traders collect the leftovers and transform them into exchangeable gifts. Even where nothing of material value to the farmers is 'stolen,' the repackaging nevertheless uses company resources and time and therefore is not necessarily encouraged by stall owners. These forms of disguised labour can be understood in terms of what de Certeau calls *"le perruque,"* the wig that masks its owner's baldness (p. 94). The appearance of carrying out legitimate work effects a *trompe l'oeil*, a trick that diverts attention away from the furtive activity that is actually taking place. These tactics allow workers to capitalize on the possibilities offered by circumstances of the moment and divert resources during the time of waged labour to something 'free' that serves the interest of individuals.

For de Certeau, the tricks and diversions of *le perruque* are individuals' ways of making do; as "tactics" (in his terminology) they do not change the formal structure or operations of an establishment. Unlike strategies, tactics do not occupy or assume a "proper space" (*espace propre*), that is, spaces recognizable as units that serve as the basis for generating relations (a stall, an enterprise, the market or an institution, for example); it is planned and administered on the basis of a number of "stable, isolatable and interconnected properties" (de Certeau, 1988, p. 94). The market might appear chaotic and messy (part of its charm and what distinguishes it from the environment of supermarkets), but is in fact a highly regulated space organized through a classificatory system of functions (storage, stall, passageways, waste, etc.) *Le perruque* succeeds when the making of timely decisions create profitable opportunities for the individual. Tactics are a "victory of space over time . . . what it wins, it does not keep" because it does not have a base where it can "capitalize on its advantages, prepare its expansions, and secure independence with respect to circumstances" (p. xix).

As such these tactics are understood to be playful, cunning and resourceful—de Certeau describes *le perruque* as "sly as a fox and twice as quick" (p. 29); its nature is often shifty, fragmentary, and elusive. Its success counts on the ability to slip between formal structures and rules and to recognize the limits upon what it can get away with; it surfs on the margins of what is permissible, teases the boundaries of that which is punishable, and tests the managers, probing the limits of their willingness to turn a blind eye. The clandestine nature of the exchanges that form the informal socio-economic spheres of the gift economy hides in plain sight, while *le perruque* lives in the interstices of the formalized mainstream.

Let us once again return to the market closing. Now there is no more commerce, the only money that changes hands are the wages being paid to the traders; managers are busy doing the final count up; workers light up their cigarettes, take off their aprons and money belts to mark the end of the workday. At this time, the mood in the market changes; to casual observers, the dimmed lights, lowered shutters, padlocks, and the empty alleyways signal the end of the market, but slowly, as more and more traders put away their aprons and gather round, another scene appears. The gutters, back lanes, and car park slowly transform as empty boxes are cleared away and crates stacked up to create a makeshift resting area. Tired legs and backs gather to joke and exchange stories of the day. Rounds of drinks are bought and lined along the closed gates of shops, gently teasing those who are still tidying up. Bags of leftover goods are passed around as final exchanges take place. Sometimes an ad hoc song is played on a guitar and empty containers turned over to serve as temporary beats to accompany the song. On sunny summer days, a game of football might take place. Tourists and passersby, no longer captured by the rustic displays of food, put their cameras away and pick their way carefully through the jungle of legs and bags lying around, waving their hands to clear the smoke in front of them.

The after-hours market opens up a social milieu in which traders could mark their presence; much like the subterranean economy of *le perruque*, these acts allow individuals to introduce their private interests into the official formal public domain. The unofficial acts of making do with the property and space of others "do not obey the law of the place, for they are not defined or identified by it" (de Certeau, 1988, p. 29) and are done so in ways that are influenced, but never wholly defined by the governing rules. Despite being on borrowed time and therefore not permanent, workers nevertheless manage to craft a place for themselves, develop a sort of familiarity, a sense of belonging, or affection for the space they were recruited to work or function in. These private acts make habitable a public space, a feeling of 'being at home.' In this sense, the reclaiming of goods and market trade space could be considered a form of tactical place-making. Place-making is not merely the taking over of a physical space, but is about the capacity to 'feel at home' in a place. Much like tenants of rented properties who mark the space with personal belongings and memories, individuals transform a foreign space into something familiar, occupying it without taking over, such that the practice of everyday life "insinuates itself onto the other's place, fragmentarily, without taking it over in its entirety" (de Certeau, 1988, pp. xii, 32). By introducing a plurality of goals and meanings to the space the effect is to "metaphorize the dominate order" (p. xx), thereby creating a habitable place for individuals that makes it function in another register.

Public/private and formal/informal binaries are no longer marked by walls or territory but by individuals' relationship to the space.

THE NEIGHBOURHOOD MARKET

The characteristics of hidden and official, or informal and formal markets outlined in this chapter do not exist exclusively or function independently of each other. The parameters of formal and informal economy; authorized and clandestine exchanges; work and leisure hours; private and public activities; gifts given and received; proper and improper spaces; mainstream and periphery are not set, nor are their relations unidirectional. The rearticulation of goods and space is not limited to the clandestine economy, nor is it limited only to the time after official business has finished. The differentiation therefore cannot be confined to the time or space to which these activities are oriented. In reality, these areas interrupt and flow together, they traverse the borders of time, place, and type of action thereby constructing an interrelated collection of socio-economic networks that is the market.

During regular trading, there are examples of goods being redirected in plain sight: giving out an extra apple or two to a regular customer, or giving a neighbouring trader a bottle of juice to start the day are common practices. This form of gift exchange does not serve only as an interruption or diversion in the formal economy but also has its own constitutive value. Customers who enjoy the recognition of their loyalty return to do their weekly shopping which benefits the shop and the farm. Neighbouring stalls return the favour with sandwiches to create a network of support and a spirit of camaraderie amongst workers. This in turn contributes to the positive atmosphere in the market that stands in contrast to the competitiveness and rivalry more often associated with markets of various kinds, from supermarkets to shopping malls or the stock market. The farmers, customers, traders, and dumpster divers all contribute to and benefit from the networks in the market which is a good source of information and support. It would be almost impossible to trace all these circuits of secondary, tertiary, and greater levels through empirical observation, as not only goods and information with primary economic function are exchanged but also intangible sentiments of affection, atmosphere, and belonging are constantly being transmitted and constructed.

Part of the reason why markets are held to be the antithesis of supermarkets or other large-scale distribution outlets is because markets are understood to give life to the neighbourhood in which they are situated. To many who keep returning, Borough Market is more than just a place where they do their grocery shopping; they feel they are part of a social environment, a neighbourhood constituted by their relationships with shopkeepers, traders, and other shoppers. Friendly exchanges, conversations, banter, and gift giving are all part of the experience, and these exchanges indicate a certain sense of belonging to a social milieu. The market is not simply a physical space but a social environment defined by the practice of shopping, shopkeeping, trading and exchanging, each individual taking up a position in the network of social relations inscribed by the environment. The parameters of the market are not defined by the topography or the surface area, rather it is *generated* by the practices of individuals inhabiting a space. A neighbourhood is therefore a social field that consists of relationships among individuals, and more precisely, according to de Certeau (1988) "the link that attaches the private to the public space" (p. 8). Hence the market could be drawn by more dynamic lines, ones that shrink and expand depending on individuals and their activities, lines criss-crossing each other in a network that gives meaning and significance to the neighbourhood.

And where does the gift exchange stand in all this? The point that Mauss makes is that the practice of gift giving "is a total system in that every item of status or of spiritual or material possession is implicated for everyone in the whole community" (Mauss, 2006, p. xi). 'Potlatch,' as the practice is known, is understood as a complete system because the relationship between members of these societies can be mapped by the catalogue of transfers which designates their obligations: "the cycling gift system *is* the society" (p. xi [my emphasis]). The importance of this integrated relationship is that gifts are used to implicate another individual, to draw another into the system of exchange it is used to strengthen relationships, create bonds and ties between tribes and clans. As Mauss points out, "a gift that does nothing to enhance solidarity is a contradiction" (p. x).

Michel de Certeau (1988) makes the similar point: an obligation to give relies upon this reciprocal, social interplay to organize and articulate a social network (p. 27). Much of the community spirit of the market is articulated in and fostered by the informalities of the after-hours market. Similar to the spirit of the Maussian gift, the care exhibited by the giving away of food and drinks articulates a certain solidarity, a sense of recognition and belonging to the market. Such an understanding situates the individual as part of a whole, hence this formulation points to the maintaining of networks of relationships, where economic dependencies, relations of patronage, and friendship are all constitutive parts of the effective functioning of the market—and hence can be argued to constitute the basic substance of the market itself.

NOTES

1. Such as Naomi Klein's *No Logo* (2001), Eric Schlosser's *Fast Food Nation* (2002), Joanna Blythman's *Shopped* (2007), Morgan Spurlock's *Super Size Me* (2004), Ruth Ozeki's *My Year of Meats* (1998), Michael Pollan's *The Omnivore's Dilemma* (2006), and Hugh Fearnley Whittingstall's TV series *Hugh's Chicken Run* (2008).
2. The *Big Issue* is a publication that is published on behalf of and sold by homeless people in the UK.

3. 'Freegans' or 'dumpster divers' refer to individuals who engage in the act of recuperating discarded food from bins. This is usually driven by an anti-consumerist ideology and not necessarily by economic needs.

4. The majority of gift giving practices in the tribes that Mauss studied are competitive in nature: the Tlinglit and the Haida tribes in the American Northwest go as far as to fight and kill chiefs in the opposing tribes to establish their status in the social hierarchy (p. 8). The act of offering gifts in these instances is often charged with hostility and rivalry. Secondly the need to outdo each other in these competitive circumstances can result in the destruction of gifts. Mauss recorded instances where the chiefs of these tribes break and throw precious copper objects and money into the rivers and seas in order to preserve their higher social status by demonstrating their superiority; for to receive without giving back is to negotiate client and servant (p. 95).

REFERENCES

Blythman, J. (2007). *Shopped: The shocking power of British supermarkets*. London: Harper Perennial.

De Certeau, M. (1988). *The practice of everyday life*. Berkeley: University of California Press.

Klein, N. (2001). *No logo*. London: Flamingo.

Mauss, M. (2006). *The gift*. London: Routledge.

Ozeki, R. (1998). *My year of meats*. New York: Viking Press.

Pollan, M. (2006). *The omnivore's dilemma: A natural history of four meals*. New York: Penguin Press.

Schlosser, E. (2002). *Fast food nation: What the all-American meal is doing to the world*. London: Penguin.

Spurlock, M. (2004). [DVD]. *Super size me*. New York: Hart Sharp Video.

Whittingstall, H. F. (2008). [DVD]. *Hugh's chicken run* (series 1). London: KEO Films.

Part II

Networks, Assemblages, and Territoriality

8   On the Beach
Informal Street Vendors and Place in Copacabana and Ipanema, Rio de Janeiro

Kirsten Seale

Copacabana and Ipanema, the renowned beaches of Rio de Janeiro, form a visual metonymy with the city that carries symbolic and real currency within interconnected cultural, spatial, and material economies. These two beaches are pivotal in the construction, communication, and regulation of Rio's place-image (Shields, 1991) as a laid-back yet dynamic pleasure-ground for the expression and consumption of local *carioca* culture. The representation of Copacabana and its more upmarket neighbour Ipanema through social media, amateur and professional photography, film, and advertising is equally crucial to experiencing the place as any embodied encounter with the extraordinary, as well as prosaic, elements of the 'Copa-nema' conjunction. This chapter explores how the presence of street markets and vendors, licensed and informal, contributes to or inhibits visual and embodied narratives of place, particularly those that present the beaches as sites of/for consumption. In doing so, it considers the discursive situating and treatment of informal street vendors along Copacabana and Ipanema beaches, and how this complex of informality, market, and place organizes or disorganizes discursive representation. This analysis is informed by fieldwork carried out in May and June 2013, along the *orla* (or waterfront precinct) at Copacabana and Ipanema, and in the suburbs proper that service the beach.

FIELD NOTES AND CONTEXT: STREET MARKETS IN RIO DE JANEIRO

Ipanema and Copacabana are located in Rio's Zona Sul, which covers the Atlantic coastal area south from Guanabara Bay to the Tijuca Massif. The Zona Sul incorporates neighbourhoods that belong to both informal and formal sectors of the city's built environment. Many of Rio's tourist attractions are concentrated here, including the Christ the Redeemer statue and the photogenic peak of Pão de Açúcar. Concomitantly, real estate in what is colloquially known as the 'asphalt city' (formal) areas of Zona Sul is some of the most expensive in Brazil.

As in the rest of Rio, and Brazil more widely, street selling was highly visible in Copacabana and Ipanema during my visit, even though the dominant groups of consumers in these areas belong to Brazil's most affluent economic demographics, Class A or B (or are employed to shop on their behalf), and are thus more likely to purchase goods from the significantly more expensive shopfronts and chain stores. Goods sold in supermarkets and chain restaurants—places of consumption frequented by the lower and middle classes in the US, Europe, and Australia—are outside the financial wherewithal of many Brazilians. To be able to afford the prices of products in shops is a mark of social status and sacrifice (Oliven & Pinheiro-Machado, 2012), and inexpensive commodities of the sort bought at informal and formal street markets are associated with the consumer practices of the poor and working classes.

In spite of movement and activity in the Zona Sul being highly regulated through municipal and privatized operations of surveillance and security, street commerce incorporating both formal and informal circuits of exchange is integrated into the local consumer ecology. Markets form a labile network of distribution and consumption, with nodes of spatially and/or temporally limited locatedness. Along many of the main shopping streets of Zona Sul are temporary and permanent market-style installations that offer everyday services and products, such as key cutting, undergarments, and snacks. There are a number of weekly produce markets that move around the Zona Sul neighbourhoods on alternate days. These take over the public squares set back a block or two from the beach and are proximate to local shopping districts. The produce markets are well frequented and are under the purview of local governance to the extent that some stalls or stands accept credit cards. In addition, there are *camelôs*, whose patches are the streets of the beach suburbs' residential and business districts, and *ambulantes*, a term used to refer to the itinerant vendors working the *orla*. Both can be licensed or unlicensed. The difference is that *camelôs* have a more or less fixed, regular pitch on a particular sidewalk, whereas *ambulantes* move throughout an area selling.

The two beachside precincts also host 'artisan' markets, which are administered by the Prefeitura de Rio (the municipal council) and service the domestic and international tourist trade with souvenirs and handicrafts. Ipanema's Sunday 'Hippie Fair' (which was, as its name suggests, set up in the late '60s as a countercultural meeting place) is a global tourist destination, to the extent that it is listed as a top shopping attraction on the TripAdvisor travel website (2013). In Copacabana, a nightly market has set up since the 1980s in the traffic island on the *orla* between the four lanes of Avenida Atlantica. There is another tourist market with similar stock further up the beach at Praça do Lido during the day. The tourist markets and the weekly neighbourhood produce markets are authorized by the Prefeitura, and through their contribution to the formal urban economy are permitted to temporarily occupy valuable space in the street. Consequently, the spatial and material infrastructure for these markets is clearly demarcated and well established.

Informal markets are a socially, if not always legally, tolerated part of the urban scene of consumption in Rio. They exist as single operators or a string of operators servicing a contingent event such as a bus queue or sporting event crowd, or a temporal event like the evening rush at the elevator which services Cantagalo on the cliffs behind Ipanema. Alternatively, unlicensed *camelôs* cluster on the fringes of the formal markets, often replicating—and undercutting—the services and products sold in the regulated areas, or providing services that are complementary or auxiliary to those offered in the market. Along the fringes of both Copacabana tourist markets, unlicensed *ambulantes* spread out further and spill over onto the beach proper. Some sell *cangas* (sarongs), which they lay out on the beach, piling sand on the corners to stop them blowing away in the breeze. As a souvenir for day trippers and holidaymakers, *cangas* are linked to the beach, materially and symbolically. They are both useful and sentimental. They are the ideal mobile commodity for buyers and sellers alike. For vendors, they are comparatively easy to transport, exhibit, and pack up—either at the end of the day, or more hastily when the authorities come around.

MARKETS AND INFORMALITY

In the Zona Sul space is a luxury commodity. According to the monthly FIPE-ZAP index (an aggregator of national real estate prices), the average cost per square metre of commercial real estate in Copacabana in June 2013 was R$10.039, up 140 per cent from January 2008. At the same time, the average cost of a square metre was R$10.000 in Ipanema, making property along the Zona Sul beaches amongst the most expensive in all Brazil. The rent-free occupation of space in these areas by unlicensed sellers is palpable. They are visibly making a living on one of the most iconic beaches in the world with no contribution to the formal economy of rent and taxes. As economically, socially, and spatially marginalized as *ambulantes* may be, they are deploying the increasingly restricted possibilities of urban public space for a livelihood.

Since Keith Hart's (1973) seminal, dualist study of informal employment in Ghana, and the coining of the phrase 'informal economy,' Diego Colleto (2010) observes that, by and large, "definitions of the informal economy are constructed in opposition to the formal or regular economy, so that the informal economy is either an intermediate stage, an obstacle, an instrumental appendage, or an alternative to the dominant model of economic development" (p. 16). For Robert Neuwirth (2011), an understanding of the informal economy which defines itself against the formal is limited in that it will always be hierarchical, with the informal positioned as the lesser, economically and ethically. At times it is conflated with criminal activity.

However, for many who work in the informal economy, discursively constructed disrepute does not align with their own perception of their work and community, nor does it take into account the vast numbers of urban dwellers capitalizing on the wide range of employment opportunities available within the informal sector (Neuwirth, 2011).

Differing terminology for the informal economy—System D (Neuwirth, 2011); grey economy; globalization from below (Mathews, Ribeiro, & Alba Vega, 2011); underground economy (Venkatesh, 2006); black market—highlights the variations in methodological and material frameworks for understanding informal economies, identifies different political and analytical agendas, and isolates varied and distinct components and attributes within what is often presented from the outside as a monolith. Informality is not only heterogeneous within. Casting the informal as the other to the formal implies that the two are in binary arrangement. However, studies of informal economies have shown the boundaries to be discursive. They do not hold up to scrutiny when one examines actualized circulations of people and goods through the permeable borders between formal and informal (Coletto, 2010, p. 35).

'Informal' connotes something that is not formed, or that lacks shape, whereas informal economies are frequently structured around a neoliberal capitalist model of *laissez-faire* free-market individualism, entrepreneurship, and self-regulation, with responsibilities and costs of employment devolved to the individual (Venkatesh, 2008). Furthermore, their growing numbers are promulgated by globalized trade, in that "less formalized economic activity operates as part of the coping strategies of people in communities caught by . . . changes in the global economy" (de Bruin & Dupuis, 2000, p. 53). In his study of workers in the informal sector in India, Jan Bremen (2013) concludes that their precarious working conditions are the direct result of the shift from rural, agricultural production to urban production as a result of globalization. Workers in the informal economy who may previously have been considered the outcasts of modernity (Bauman, 2004) are now the embodiment of a neoliberal politics that coalesces around the primary characteristics of contingency, mobility, and flexibility in global labour. Workers in the informal economy are therefore a core set of workers in the global economy. They are hardly marginal, even though they might be marginalized economically and politically.

The informal economy is, above all, an active response to the incapacity of the private and public sectors to provide employment. An International Labour Organization (ILO) report from 1972, also known as the 'Kenya Report,' defined the informal sector as a series of economic activities characterized by "ease of entry; reliance on indigenous resources; family ownership of enterprises; small scale of operation; labor-intensive and adapted technology; skill acquired outside of the formal school system; and unregulated and competitive markets. [They are] largely ignored, rarely supported, often regulated and sometimes actively discouraged by the Government"

(International Labour Office, 1972, p. 6). On the other hand, the report found that the financial and technological imposts, and the social obstacles of setting up a business overseen by the state are outside the capacity of many workers in informal economies, and therefore function as an effective disincentive to operating within the state's purview.

Formal urban economies, such as the network of official markets in Copacabana and Ipanema, construct and regulate flows of bodies, labour, and capital through their placement in urban space. By contrast, the movement, and the stasis, of informal street markets and vendors in the streets are *responsive* to flows of movement, people, goods, and communication in urban space (see Hepworth in this volume). On Ipanema, I watched a swimwear salesman who had pinned his costumes to a circular frame that he held above his head like a parasol as he walked up and down the sand. It had become a kinetic display device, swaying colourfully. It moved, and could be moved. Despite the physically demanding and competitive work of walking the beach in the heat with his load, he stopped to chat and call out greetings to rival sellers, and to the workers at the *barracas* hiring out umbrellas and chairs. His path was never predetermined or prescribed, except by the parameters of the beach and its associated architecture.

Figure 8.1   Ipanema Beach.

I saw other beach sellers, licensed and unlicensed, with a similarly relaxed demeanour. It appeared at odds with the precarious and demanding circumstances of their work. Colleto (2010) interprets this as a distinctively Brazilian aptitude for informal labour.

> [O]ne gains the distinct impression on observing the downtown streets of the large South American cities that the freedom of the individual is realized more fully in those spaces than in other contexts. . . . In the Brazilian culture, this aptitude is expressed by the term *jeito*. . . . Behaving with a certain *jeitinho* . . . is not just to improvise mechanically but . . . an aptitude that seems crucial for survival in informal street trade.
>
> (p. 101)

Aside from a perception that Brazilians are uniquely equipped to take on such material conditions, and the acknowledgement that such labour demonstrably provides relief from social and economic exclusion, Colleto is nevertheless mindful that it can also be what Portes and Roberts call "forced entrepreneurialism" for workers, which eventuates from political conditions where "jobs are not available, or are of such poor quality as to keep those holding them in permanent poverty" (2005, p. 49). Similarly, while Ananya Roy (2011) argues in favour of "subaltern urbanism, . . . an important paradigm, for it seeks to confer recognition on spaces of poverty and forms of popular agency that often remain invisible and neglected" (p. 224)—a paradigm and recognition which surely encompasses informal street selling—at the same time she acknowledges that globalization from below is not always as equitable as it sounds.

PLACE-IMAGE ALONG THE BEACH

Given the prevalence of informal markets and markets at Copacabana and Ipanema, and consumers' apparent ease with their presence on the beach (played out through phatic and commercial exchange), it could be surmised that informal markets and vendors are part of the place-image of Copacabana and Ipanema. Informal street selling is a grounded, everyday practice; unexceptional in spite of governmental policies and operations that attempt to exceptionalize it. In his account of how place-image is discursively constructed and transmitted, Rob Shields (1991) cites the example of the beach.

> Through a process of labelling, sites and zones associated with particular activities become characterised as being appropriate for exactly those types of activities. Other activities are excluded, forced into the wilderness or barren spaces "outside" of civilised realm, or they are associated with their own dichotomous spaces. . . .

[Place-images] are the various discrete meanings associated with real places or regions regardless of their character in reality. Images, being partial and often either exaggerated or understated, may be accurate or inaccurate. They result from stereotyping, which over-simplifies groups of places with a region, or prejudices towards places or their inhabitants. A set of core images forms a widely disseminated and commonly held set of images of a place or space. These form a relatively stable group of ideas in currency, reinforced by their communication value as conventions circulating in a discursive economy.

(p. 60)

Real estate along the Zona Sul beaches may be some of the most expensive in Brazil, but, as Godfrey and Arguinzoni (2012) have discussed, the dominant place-image of the beaches themselves, within Brazil and globally, is of pluralistic and socially accessible spaces that are representative of the inclusiveness of Brazilian society. Even so, this representation is contested and countered by the signals and instruments of another narrative or place-image along the *orla*. In 2010, tents appeared along the beach to support and promote *Choque de Ordem* (Shock of Order), the city's militarily enhanced policy of bringing "pacification" to the *favela* neighbourhoods behind the beaches. It was no spatial coincidence that the Zona Sul *barrios* of Morro da Babilônia/Chapéu-Mangueira and Cantagalo-Pavão-Pavãozinho, adjacent to the valuable real estate of Copacabana and Ipanema, were among the first to be occupied by the UPP (Police Pacification Units) in the ongoing militarized occupation of Rio's *favela* neighbourhoods since 2009. Part of the citywide *Choque* strategy stretched to the Zona Sul beaches in the form of a crackdown on unlicensed vending. Through the control of flows of people and goods along the beach, or at least giving the impression that this had been achieved, the Prefeitura's objective is a place-image of a beach secure from informal activity, which in this case is frequently discursively conflated with criminality.

This strategy is reflected and refracted in the built environment along the beach. Newly constructed kiosks on Copacabana's beachfront have replaced the more shack-like bars and their ramshackle plastic chairs and red and yellow umbrellas. The new glass and steel structures are more permanent and 'café-like,' with semi–cordoned off, undercover eating areas. The barriers are both material and social, protecting the clientele from the *ambulantes* and panhandlers who roam the beach precinct. The kiosks are part of the current 'clean-up' Copacabana is undergoing as one of Rio's four designated 'Olympic Zones' at the 2016 games. This renewal program is merely the latest iteration in a series of officially constructed place-images for Copacabana in the 120 years of its settlement. The beach precincts of Copacabana and Ipanema are products of a 20th-century, modern notion of the beach (Lencek & Bosker, 1998), which viewed the beach as the stage for recreation and pleasure. Copacabana, since its inception, has always been a site that signified the dissolution of social inhibitions for those with the time and finances to afford them (Castro, 2004). In more recent times, the beach as place-image has expanded to signify a pluralistic and polyglot utopia in a conspicuously socially and spatially segregated city (Godfrey & Arguinzoni, 2012).

Amongst the many *cariocas* and tourists, who are walking, cycling, sitting with an *agua de coco*, playing beach volleyball or *frescobol*, there are vendors who work at the *barracas* and the *quiosques*. There are also licensed and unlicensed *ambulantes* who sell favourite local snacks and beverages, such as *mate*, *açai*, Globo crisps, sandwiches, and peanuts as well as global brands of soft drink and ice cream. Some vendors sell site-specific products—swimming costumes, resort wear, sun hats, and sun cream—whilst others sell cigarettes, temporary tattoos, and silver and beaded jewellery. The emphasis on democratically glamorous consumption in the visual and material economies of the beaches brings into relief the division between those who work on the beach and those who do not. The presence of some sellers who are visible reminders of "advanced marginality" (Wacquant, 2008) causes a rupture in the smoothness of any unilateral meaning of the beach as an egalitarian space of/for consumption, and reinscribes the tensions "between social order, status, and hierarchy on one hand, and democratic rights, social diversity, and accessibility on the other" (Godfrey & Arguinzoni, 2012, p. 18).

Figure 8.2   Ipanema Beach.

URBAN HETEROTOPIAS

For anthropologist Roberto da Matta (1991), the visitors to the newly constructed, partially fenced-off kiosks along Copacabana are acting out what
he has called "the Brazilian dilemma." According to da Matta, this is a series
of binaries in the Brazilian collective imagination, one of which stigmatizes
the street as the dangerous and unpredictable other to the home. The street
is the sphere of "punishment, struggle, and work," and is conceived "in
Hobbesian terms: it is the way of all against all until some form of hierarchical principle can surface and establish some kind of order" (p. 65). Given
the overlap between the street and the beach in terms of their discursive
situating within public space and the public imagination, it could be that the
beach is as susceptible to collective anxieties about social order and disorder.
The *Choque de Ordem* tents on the sand at Copacabana and Ipanema were
a reminder of moral panic surrounding the use and practice of public spaces.

As Benjamin Shepard and Gregory Smithsimon (2011) elaborate, not everyone perceives the street, or the beach, this way. One of the resonating slogans
of the revolutionary urbanism of the May 1968 protests in Paris was "*Sous les
paves, la plage*," or "Beneath the pavement, the beach." Shepard and Smithsimon (2011) conclude that the beach in "The anonymous graffiti . . . conjures
up any number of images—a subaltern vitality, the control of something unruly,
the dominance of nature and a possible return of the repressed. The expression
also speaks to a new kind of social imagination, a right to view the city as a
space of democratic possibilities, a social geography of freedom" (p. 3).

For informal street vendors, the street can present possibilities for livelihood and community, for the creation of the "subaltern urbanism" that
Roy (2011) advocates. Others conceive of the urban, and the market in
particular, as a heterotopia that is both the incubator and the outcome of
intersections with the other. In her research on markets in London and the
UK, Sophie Watson (2006; 2009) documents how heterotopic encounters
with the other, in the form of "rubbing along," not only produce sociality
and conviviality in the urban street market, but are crucial to the market's
feasibility as public space. Michel Foucault (1986) writes of heterotopia as
a spatial phenomenon "capable of juxtaposing in a single real place several
spaces, several sites that are in themselves incompatible" (p. 25). He also
stresses a temporal dimension to the heterotopic.

There are those that linked to time in its most flowing, transitory, precarious aspect, to time in the mode of the festival. These heterotopias
are not oriented toward the eternal, they are rather absolutely temporal
[*chroniques*]. Such, for example, are the fairgrounds, these marvellous
empty sites on the outskirts of cities that teem once or twice a year with
stands, displays, heteroclite objects, wrestlers, snakewomen, fortune-
tellers, and so forth.

(p. 26)

Meaghan Morris (1998), too, urges us to read the beach in terms of the temporal, along with the spatial. She deploys the Bakhtinian "chronotope" to achieve this.

> *In this spirit*, it may be more useful to think of the beach as a *chronotope* rather than as a *topos* or myth. Bakhtin's famous "unit of analysis" based on variable time-space rations can carry its own essentialist charge, but it does allow us to deal with the density and volatility of cultural reference systems without . . . bringing an impossible totality relentlessly to bear on every single occasion . . . The *beauty* of the concept of *chronotope* is to enable us to think about the cultural interdependence of spatial and temporal categories in terms of *variable* relations.
> (pp. 104–105)

To do as Morris suggests, and transpose the chronotope from the field of literature, where Mikhail Bakhtin situated it (1986, p. 84), for use as a tool to understand the beach at Copacabana or Ipanema, recognizes that synchronous and/or competing narratives of place emerge from continually reorganizing past, present, and future entanglements (Ingold, 2011) of market, vendor, consumer, informality, place, and the beach/street. Shields'

Figure 8.3   Ipanema Beach.

less dynamic notion of place-image as constellation of narrative, representation and practice acknowledges the discursive element of this "meshwork of entangled lines of life, growth and movement" (Ingold, 2011, p. 63). Moreover, the discursive process Shields describes at work with place-image shares some of the characteristics of the emergence of the spatial/temporal phenomenon of heterotopia as articulated by Foucault: "As a sort of simultaneously mythic and real contestation of the space in which we live, this description could be called heterotopology" (Foucault, 1986, p. 24).

In the popular imagination, few urban spaces are more resonant of local place than the street market. At the same time, markets have historically functioned metonymically in relation to the heterogeneity of identities, encounters, and practices that constitute the urban. They are indexical to a "right to the city" (Lefebvre, 1996). In the face of the hegemonic instruments of automobility, surveillance technology, and totalizing spatialities such as urban renewal, the market reclaims the potential of the street as a heterotopic space in the Foucaultian sense (1986; 2002), and in doing so, actualizes the market's historic condition (Stallybrass & White, 1986) as an ambivalent place beyond category—somewhere between the formal and informal, the here and elsewhere, the global and local. The Zona Sul beaches can also be experienced as an urban heterotopia. In Rio de Janeiro, challenges to the heterotopic informality produced by the street market and street vendors, particularly those that use the beach as their stage, are presented by the interconnected forces of construction and urban planning for two megaevents (the 2014 FIFA World Cup and the 2016 Olympic Games), as well as moves towards a global economy, and an increase in property values. In spite of the legal, political, and media discourses supporting these forces that stigmatize and in some cases criminalize street commerce, the unlicensed vendors at Copacabana and Ipanema are very much part of the place-image of the beach as scene of consumption, tourism, and the everyday.

ACKNOWLEDGEMENTS

Thank you to Nicholas Mariette and David Seale for helping make this research happen.

REFERENCES

Bakhtin, M. (1986). Forms of time and of the chronotope in the novel (V. W. McGee, Trans.). In C. Emerson & M. Holquist (Eds.), *Speech genres and other late essays* (pp. 84–258). Austin: University of Texas Press.

Bauman, Z. (2004). *Wasted lives*. Cambridge, UK: Polity.

Breman, J. (2013). *At work in the informal economy of India: A perspective from the bottom up*. Oxford: Oxford University Press.

Castro, R. (2004). *Rio de Janeiro: Carnival under fire*. London: Bloomsbury.

Coletto, D. (2010). *The informal economy and employment in Brazil: Latin America, modernization, and social changes.* London: Palgrave Macmillan.

Da Matta, R. (1991). *Carnivals, rogues, and heroes. An interpretation of the Brazilian dilemma* (J. Drury, Trans.). Notre Dame, IN: University of Notre Dame Press.

De Bruin, A., & Dupuis, A. (2000). The dynamics of New Zealand's largest street market: The Otara Flea Market. *International Journal of Sociology and Social Policy, 20*(1–2), 52–73.

Foucault, M. (2002). *The order of things: An archaeology of the human sciences* (A. Sheridan, Trans.). London: Routledge.

Foucault, M., & Miskowiec, J. (1986). Of other spaces. *Diacritics, 16*(1), 22–27.

Godfrey, B. J., & Arguinzoni, O. M. (2012). Regulating public space on the beach-fronts of Rio de Janeiro. *Geographical Review, 102*(1), 17–34.

Hart, K. (1973). Informal income opportunities and urban employment in Ghana. *Journal of Modern African Studies, 11*, 61–89.

Ingold, T. (2011). *Being alive: Essays on movement, knowledge and description.* Hoboken, NJ: Taylor & Francis.

International Labour Office. (1972). Employment, incomes and equality: A strategy for increasing productive employment in Kenya.

Lefebvre, H. (1996). *Writings on cities* (E. Kofman & E. Lebas, Trans.). Oxford: Blackwell.

Lencek, M., & Bosker, G. (1998). *The beach: A history of paradise on earth.* London: Secker & Warburg.

Mathews, G., Ribeiro, G. L., & Alba Vega, C. (2012). *Globalization from below: The world's other economy.* London: Routledge.

Morris, M. (1998). *Too soon too late: History in popular culture.* Bloomington: Indiana University Press.

Neuwirth, R. (2011). *Stealth of nations: The global rise of the informal economy.* New York: Anchor.

Oliven, R. G., & Pinheiro-Machado, R. (2012). From "country of the future" to emergent country: Popular consumption in Brazil. In J. Sinclair & A. C. Pertierra (Eds.), *Consumer culture in Latin America* (pp. 53–65). New York: Palgrave Macmillan.

Roy, A. (2011). Slumdog cities: Rethinking subaltern urbanism. *International Journal of Urban and Regional Research, 35*(2), 223–238.

Shepard, B. H., & Smithsimon, G. (2011). The beach beneath the streets: Contesting New York City's public spaces. Albany, NY: Excelsior.

Shields, R. (1991). *Places on the margin: Alternative geographies of modernity.* London: Routledge.

Stallybrass, P., & White, A. (1986). *The politics and poetics of transgression.* Ithaca, NY: Cornell University Press.

Tripadvisor. (2013). Hippie Fair Crafts Market. Retrieved from www.tripadvisor.com.au/Attraction_Review-g303506-d317891-Reviews-Hippie_Fair_Crafts_Market-Rio_de_Janeiro_State_of_Rio_de_Janeiro.html

Venkatesh, S. A. (2006). *Off the books: The underground economy of the urban poor.* Cambridge, MA: Harvard University Press.

Venkatesh, S. A. (2008). *Gang leader for a day: A rogue sociologist crosses the line.* London: Allen Lane.

Wacquant, L. (2008). *Urban outcasts: A comparative sociology of advanced marginality.* Cambridge, UK: Polity.

Watson, S. (2006). Nostalgia at work: Living with difference in a London street market. *City publics: The (dis)enchantments of urban encounters* (pp. 41–61). London: Routledge.

Watson, S. (2009). The magic of the marketplace: Sociality in a neglected public space. *Urban Studies, 46*(8), 1577–1591.

9 Pengpu Night Market

Informal Urban Street Markets as More-Than-Human Assemblages in Shanghai

Clifton Evers

Tricycles (*San Lun Che*) are common on the streets of China's cities, and part of their informal urban street markets. In this chapter I focus on several 'becoming-tricycle assemblages' on the streets of Shanghai, China.

By using the term 'becoming-tricycle' I am following the collaborative work of philosophers Gilles Deleuze and Félix Guattari (1987) who explain that 'becoming' refers to an ongoing process of collecting, discarding, interacting, and blocking. This is a move away from thinking about tricycles (or informal urban street markets in general) as objects towards interrogating them as what Deleuze and Guattari refer to as "assemblages." Assemblages are an interplay of relationships *between* heterogeneous elements that bring out particular capacities of those elements and that yield 'articulations' because of those relationships. By 'articulations' I mean properties, tendencies, momentums, forces, qualities, meanings, and behaviours (Anderson & MacFarlane, 2011).[1]

I understand each becoming-tricycle as a mobile, informal urban street market assemblage in and of itself, with the ability to become as part of other such assemblages of various scales and intensities. These assemblages are only fragments of happenings in a rapidly growing city of over 23 million people (Shanghai Municipal Information Office & Shanghai Municipal Statistics Bureau, 2011).[2]

My focus is on how becoming-tricycle played out in relation to the Pengpu night market, a larger scale assemblage. Through this empirical study I register the force and energy of informality "in its own terms" (Marx, 2009, p. 337). For me, informality in its own terms equates to detours, derailments, hijackings, discoveries, arbitrariness, leading astray, cutting up, refusing, collaging, appropriations, experimentations—of regulations, bodies, space, materiality, and subjectivity. I argue that such informality produces "ambiances" (Khatib, 1958) that enable livelihoods to emerge in Shanghai. By 'livelihoods' I mean interrelated and co-constituted financial, human, natural, physical, and social actions.

In 1958, Abdelhafid Khatib stressed the importance of the ambiance of street markets while undertaking a "psychogeography"[3] of the Les Halles district in Paris, France. He described how a "logjam of lorries, the barricades of panniers, the movement of workers with their mechanical or hand barrows, prevents access to cars and almost constantly obliges the pedestrian to alter his route" (n.p.). Khatib notes that this arrangement is temporary and variable by the hour. They result in what he refers to as "ambiances"— a commingling of sensations, sounds, smells, atmosphere, actions, space, architectures, and bodies, among other elements.

I want to emphasize the role of the nonhuman, just as well as the human, during informal urban street markets when ambiances emerge. Ambiances are "not just willed by us humans but come about equally through the materialities of the world in which we are just a part" (Dewsbury, 2011, p. 152). As Alphonso Lingis (1994, p. 96) argues, there is a "murmur of the world" that is the "echo of the vibrancy of things. To be, for material things, is to resonate" as warmth, sound, and affect. There is a vitality, vibrancy, and efficacy to matter (Bennett, 2010).

While informal urban street market assemblages such as becoming-tricycle and the Pengpu night market achieve a sufficient resonance, coherence, intensity, and direction to reach a provisional state of durable equilibrium for a few hours each evening, they are always in a state of "becoming" (Deleuze & Guattari, 1987). The movement, complexity, and open-endedness of this "becoming" is important because it draws our attention to what Deleuze and Guattari (1987) refer to as a "micro-politics." This is a politics that is an ongoing subversion of meta-narratives, regulations, state apparatuses, conventions, imposed subjectivities, formalization, and standardization, and thus a possible liberation from such, and their tendencies for producing hierarchical and subsequently inequitable and alienating outcomes. This inspires a mode of thinking about how the various elements of assemblages get along (or don't), how various relationships play out, how articulations manifest (or are blocked), how new modes of being can emerge, and how this mode of thinking—this micro-politics—itself is vitalized by the more-than-human dynamism of informal urban street markets.

The central questions in this chapter are: What does 'becoming-tricycle' do? How does this process make a difference and alter, enable, and produce informal urban street markets and their concomitant micro-political ambiances in Shanghai? And what propositions, possibilities, and considerations are brought forth by paying attention to such?

* * *

Pengpu night market was located in the Zhabei district in Shanghai. It began in 2004 when the Pengpu Xincun Metro station on Line 1 became operational. It was closed in December 2013. While the night market had begun small by late 2013, it had swelled to approximately 400 stalls and stretched one kilometre along Linfen Road. Buses had to make detours from their official routes to avoid the congestion (Jian, 2013). The market was made up of stalls selling food, home ware, clothing, and much more. The main draw

Figure 9.1   Tricycles as assemblage.

was the street food. Tricycles were part of this assemblage, which involved more-than-human agency and affinities.

The tricycles that converged at the market were in various states of disrepair. Many of them were held together by tape and amateur metallurgy. The metal had rusted and broken due to chemical activity. Metal has a vitalism of its own (Bennett, 2010). Torn and tattered tarpaulins were folded and ready if rain demanded them. Tricycles became stalls with a vast array of goods—DVDs, handbags, clothing, and technologies—that ignored intellectual property laws: Apple becomes Appel, Versace becomes Versase, Gucci becomes Guchi. Customers who normally would be excluded from the cultural capital economy of such goods got to take part. Also being sold were fruit and vegetables, books, crockery, and household goods. One street trader had a tricycle with a machine built into it that crushed long shafts of sugarcane into a sweet drink—a sugar-cane-crushing-juicer-tricycle. Many street traders had a gas bottle under the tricycle tray and cooked food in woks, all the while being surrounded by small plastic chairs—the tricycle as stove-restaurant. Vendors would negotiate among themselves locations along the street. Some would settle into a position and claim it as their own, while others would move about. They could come and go at will. Mobility was enabled by these tricycles. In the mornings, rats and birds would feed off the waste until the local government cleaners arrived. The more-than-human relationships produce the possibilities of the market.

While I have come to understand becoming-tricycle as assemblage, the Pengpu night market more broadly was an assemblage too. Tricycles, street vendors, stalls, trees, buildings, customers, economics, policies, institutions, city transport infrastructure, and so on drew each other in, functioning together and co-constituting the emergence each night of a provisional informal urban street market assemblage. The intensity and size of this informal urban street market assemblage waxed and waned. It was ever-moving, shape-shifting, and self-organizing. Customers were created as tricycles, people, produce, and the like moved about. Agential influence manifested out of the relationships in various forms. That is, not simply from one particular element but from the 'in-between'—the liminal.

An ambiance emerged that drew in customers (locals and tourists), and so enabled capital and sociality to flow, and livelihoods to emerge. I was drawn in. When I walked through the markets my senses would be overloaded. There was intensity to the ambiance that was felt and negotiated. People bumped into me. I slipped on food waste and oil on the ground. I would confuse voices with the ring of a bell. I got frustrated with the crowds. I experienced joy when my friends arrived and we would sit down on the tray of a tricycle and eat together. We would be swept up by the ambiance. My sense of self would blur into the commingling—the ambiance was all that there was.

There was a registering by the body-as-human that something more-than-human was happening and important during ambiances of Shanghai. This was an "intertwining of body and world as a set of intensities" and the blurring of thresholds, such as the "boundary between body and environment" (Jones, 2012, p. 645). This more-than-human orientation reminded me of how Elizabeth Grosz (2002) critiques the idea that bodies and cities are simply involved in a binary of cause and effect, that cities are simply products of human needs, or conversely, cities are 'unnatural' and alienating to bodies. Rather, Grosz proposes that bodies and cities are co-constitutive, mutually defining, and productive. However, as a *laowai* (alien), my perspective could be said to be a romanticization and exoticization of the ambiance that took place. As several street vendors explained to me, livelihoods that emerge from becoming-tricycle involve *chiku* (eating bitterness) (Loyalka, 2012). Winter can get bitterly cold, competition for customers can be fierce, the profits generated can be meagre, and authorities can make life difficult. Yet this is a market frequented by mostly locals. And when I asked some vendors if they would prefer to do something else they said "no," they enjoy the "vending life" despite the difficulties.

It has been argued that such street vending causes traffic problems, is 'unsightly,' and that its informality interrupts the functioning of "legitimate businesses" (Chai et al., 2011). In some places in the world it is not uncommon for informal street markets to be "known for criminal activity, from theft and other forms of petty crime to the sale of illegal goods and services," even though there may be no such activity taking place (Neuwirth, 2011, p. 21).[4] In China this seems to be less the case (Chai et al., 2011). Even so, some authorities believe street vendors are costly because they provide little direct revenue for governments, considering the cost in regulating them.

Over the years there were a number of crackdowns by the local Shanghai government to check that vendors operating at the Pengpu night market had approved licences to operate (purchased from that same local government). The majority of vendors did not have licences or pay any taxes or fees. There were arguments and sometimes even fighting when authorities confiscated goods, stalls, and tricycles. During November 2013 there were a series of aggressive overnight crackdowns by the local government. It was argued that the market had become too large and disruptive, for example interrupting bus services. Some residents of Linfen Street had complained for years about the noise, smoke, sanitation, traffic congestion, and disorder of Pengpu night market. One resident wrote in an article in the local press: "I hope this market is gone for good so that we can resume our untroubled life in this community the way it was 10 years ago" (Zhenqi, 2013). Another reason for the crackdown was the emphasis on the modernization and increasing commercialization of the Zhabei district where the Pengpu night market was located, and Shanghai more broadly.

Further, the assemblage was problematizing any clear division between public and private space by ignoring rents, leases, and freeholds. The assemblage had enabled a provisional territory to emerge. As Wise (2005), following Deleuze and Guattari, points out:

> territories are more than just spaces; they have a stake, a claim . . . Territories are not fixed for all time, but are always being made and unmade, reterritorializing and deterritorializing . . . always coming together and moving apart.
>
> (p. 79)

What emerge are relationships between and negotiations of "striated space" and "smooth space" (Deleuze & Guattari, 1987). Striated space is ordered, systematic, restricted, and closed. Smooth space is heterogeneous and transgressive, a multiplicity of diverse connections, encounters, and becomings. These are not opposed models but rather "the two spaces in fact coexist only in mixture: smooth space is constantly being translated, traversed into striated space; striated space is constantly being reversed, returned to smooth space" (p. 23). It is a process "perpetually in construction or collapsing" (p. 22).

Police and *chengguan* ('urban management officers') did the enforcing of striated space.[5] *Chengguan* have no legal authority, but are used by the government to augment police in enforcing municipal laws, regulations, and codes. The crackdowns were particularly worrisome for vendors without a

Shanghai *hukou* card—a place-specific urban residency card connected to a registration system. Some of the street vendors who were becoming-tricycle at the Pengpu night market were migrant workers (*mingong*). They arrive from rural or regional areas outside Shanghai. Migrants move to Shanghai from provinces such as Anhui, Jiangsu, Henan, and Sichuan. Urbanization has involved a large population movement (*liudong renkou*) to the major cities (Meng et al., 2010). Nationally, in 2011 there were at least 262.1 million migrant workers (Shanghai Municipal Information Office, 2012). The flow of people became regulated because of the pressures it brought to bear on city space and resources, as well as government services, planning, and finances. The regulation (blockage) to the flow of people occurs via the resident registration system (*hukou*) (Cai, 2000; Meng et al., 2010). Without a place-specific urban residency card the migrants have only limited access to various social services, such as schools, health care, work, and pension insurance (Meng et al., 2010).[6] As Dianne Currier (2003) points out, assemblages occur in relation to "regimes of signs and relations of power," which aim to and can "achieve a meta-stability" (p. 327).

People move to Shanghai to find employment, although some of the younger people also seek adventure and to escape the 'boredom' of village life, as one young vendor told me. The migrants use kin (*xueyuan*) and place-ties (*diyuan*) to raise money and purchase their tricycle and goods to sell, or to make adaptations to their tricycles in order to offer services. Some get their start-up capital by borrowing from the pooled savings (*jīxù*) of their working group (*dānwèi*). Informal banking is common in China (Tsai, 2002). Becoming-tricycle and the concomitant street vending have low barriers to entry in terms of cost and education, which makes it possible for people to generate livelihoods for themselves. Some people choose vending owing to its flexible work hours and for social and cultural reasons, such as gender, interaction, physicality, and tradition. Becoming-tricycle provides opportunities for livelihoods over which people may have more control; for example, it provides working conditions that are flexible for those with a high degree of familial duties. Also, becoming-tricycle is a way of avoiding the somewhat Kafkaesque red tape and bureaucracy of the Chinese government.

Becoming-tricycle allowed some people to become part of the night market when it was 'smoothed out' but to move away as 'striation' became dominant, when the authorities moved in. Becoming-tricycle articulated livelihoods that were underscored by vigilance, adaptation, creativity, and mobility. While these livelihoods can be secondary, seasonal, temporary, or part-time, they can, for others, be the primary occupation and source of income. So, fines or confiscation of goods and tricycles creates significant problems. When the authorities would arrive during the crackdowns some vendors would hurriedly convert their tricycle-restaurants to transport and move away, hoping to be ignored by the authorities. Without a Shanghai *hukou* card they could be expelled from the city. Without a licence they could be fined. They would leave the night market, only to return later. This strategy of cat and mouse with various authorities and the attendant formal rules, policies, and regulations have become a familiar part of the becoming-tricycle-informal-urban-street-market assemblage in Shanghai.

When vendors moved away from Pengpu night market some would travel to other locations in the district (or even to other districts). They would cycle through the thick pollution of Shanghai, lungs burning as sulphur dioxide, nitrogen dioxide, and carbon monoxide changed their lung tissue. They had to negotiate the heavy traffic of Shanghai. Painted white lines on the bitumen and tri-colour traffic lights meant to regulate traffic flow would act only a guide and not a rule. The traffic flow was somewhat managed but also self-organizing with a cacophony of horns sounding to signal where others are, distances between bodies, gaps appearing, blockages coming up, and traffic rules to break.

Becoming-tricycle involves a process of "enskilment" (Pálsson, 1994)—a situated, affective and contextual learning. Enskilment occurs where senses and the more-than-human work to move beyond and undermine any subject–object nature–culture interpretation and experience of subjectivity. Becoming-tricycle involves a heterogeneous assembling of arms, legs, flesh, metal, wind, jacket, wheels, grease, rubber, handlebars, metal, space, ropes, sensations, regulations, sounds, and teetering.

Two vendors temporarily set up on a corner near my home for a few weeks following one crackdown. People would stop to eat. A flute player would arrive, lay out a rug, and play his music. Passers-by placed coins on the rug. Another tricycle arrived one night, loaded with home ware. The footpath became impassable and people began to spill onto the street to get by. Vehicles were forced to slow down when the assemblage took place, most evenings after 5:00 p.m. An ambiance emerged and dissipated each evening. I had several conversations with the vendors. They viewed themselves as legitimate businesses (Chai et al., 2011). They explained that they provide valuable goods and services that are convenient, affordable, and accessible. For example, most of the 200,000 residents in the Pengpu community are relocated residents from old torn-down neighbourhoods and non-Shanghainese who have relative low buying power, making the night market with its cheap food and products extremely popular. Further, the street market provided commodities in small quantities and when permanent stores had closed for the evening. One of the vendors explained how she cooks for some families whose residences have no kitchen, a common situation in some districts of Chinese cities (Zhang, 2001). Despite all these features, in December 2013 the Pengpu night market was permanently shut down. Barriers were erected. *Chengguan* monitor the area. But the vendors move about the city again and becoming-tricycle continues.

***

By exploring becoming-tricycle and the Pengpu night market I have come to realize that informal urban street markets are what Kim Dovey (2012) calls "complex adaptive assemblages." These assemblages are mobile, multisca-lar, provisional, spatial, and constituted across various registers—chemical, biological, psychological, and social.

These complex adaptive assemblages produce ambiances that feed off informality. There are many life-sustaining and producing quotidian tactics taking place during such more-than-human-informal-urban-street-market-assemblages—such as becoming-tricycle. Yet, there continues to be insufficient attention given to accommodating these during the rush to modernization in Shanghai. Further, there is insufficient attention given over to what such ambiances and informality can reveal about alternative conceptions of space, sociality, economics, bodies, and materiality. That is, how they can enable us to register "the sound of a contiguous future, the murmur of new assemblages" (Deleuze & Guattari, 1987, p. 83).

I have drawn attention to the marginalization or repression of informality by state power (for example, district governments) through the regulating and ordering of public spaces, flows of people, and access to services through ratio-nal planning. I have also noted some modes and strategies of resistance to such impositions the street vendors employ and the assemblages produce. How-ever, what has come to significantly stand out for me through this research is how informality, creativity, and becoming are interwoven—the micro-politics. Such a more-than-human micro-politics can, following Ananya Roy (2011), "disrupt models of expertise" and create room for alternative forms of knowl-edge and living to come forth that are beyond the purview of planning, that is the "unplannable." The research confirms Roy's point that informality produces "the ever-shifting urban relationship between the legal and illegal, legitimate and illegitimate, authorized and unauthorized" (p. 233).

The subsequent challenge, then, is a critical reexamination of any approach, planning, policy, or regulation that simply aims to contain, disentangle, systematize, order, control, discipline, 'clean up,' manage, pathologise, criminalize, and penalize informal urban street markets, and the people for whom these are necessary. For when we do such certain hege-monic power relations and hierarchical organization are reinscribed. This may broaden how we value informal urban street markets in terms of what they do and produce as alternative articulations of subjectivity, identity, bodies, space, meanings, regulations, and so on. There is a need to listen to and learn from such nonhegemonic modes of becoming-city in this age of rapid urbanization if we want to ensure that the ambiances of cities mean liveability and livelihoods for more than a few.

ACKNOWLEDGEMENTS

This chapter came out of work done with the Community Markets Project, supported by RMIT and the University of Nottingham Ningbo China. I would like to thank Katarina Olausson for her valuable feedback on the drafts and Wei Song for her help with some interpretation and translation.

NOTES

1. However, it is important to note that "an assemblage does not proceed by way of distinct unities coming to bear on each other within an already established framework, but on entities and forms, discourses and institutions, achieving mutual and localized constitution and becoming operational within the context of the particular assemblage within which they are articulated" (Currier 2003, p. 327).
2. This figure refers to local and migrant populations living in Shanghai.
3. Psychogeography was originally developed by the avant-garde movement Lettrist International in the journal *Potlach*. Psychogeography and one of its methods *dérive* (calculated drifting through the urban environment) were first developed by Ivan Chtcheglov, in his 1953 essay "Formulaire pour un urbanisme nouveau" (Formulary for a New Urbanism). This Lettrist International was a collective of European artists and theorists, a precursor to The Situationist International group. The Situationist International artist/thinker/activist Guy Debord (1955) clarified and developed called psychogeography "a whole toy box full of playful, inventive strategies for exploring cities . . . just about anything that takes pedestrians off their predictable paths and jolts them into a new awareness of the urban landscape" (p. 8). Khatib was a member of The Situationist International collective.
4. It is important to note that informality does not simply equal underdevelopment, illegality, marginality, or poverty (Dovey, 2012, p. 352). The informal economy is part of and even central to some of the most developed sectors of societies and economies (Portes, Castells, & Benton, 1989; Dovey, 2012). For example the informal sector also includes informal information technology creative clusters in advanced economies.
5. *Chengguan* have a poor reputation and are known for confiscating goods, as well as thuggish behaviour, such as fining (extorting), intimidating, harassing, manhandling, and assaulting vendors, which has sometimes resulted in deaths and riots (Human Rights Watch, 2012). There is a formality to the *chengguan* (e.g. rules and procedures meant to be followed); however, this is entangled with informality as some *chengguan* act outside that formality or recalibrate the role for personal gain as they struggle to survive on low wages and establish a livelihood.
6. Some rural to urban migrants can eventually obtain a permanent urban *hukou* status.

REFERENCES

Anderson, B., & McFarlane, C. (2011). Assemblage and geography. *AREA, 43*(2), 124–127.

Bennett, J. (2010). *Vibrant matter: A political ecology of things.* Durham, NC: Duke University Press.

Cai, F. (2000). The invisible hand and visible feet: Internal migration in China. *World Economy and China*, Working Paper No. 5. Available at www.rrojasdatabank. info/caifang2000.pdf

Chai, X., Ziqiang, Q., Pan, K., Deng, X., & Zhou, Y. (2011). Research on the management of urban unlicensed mobile street vendors: Taking public satisfied degree as value orientation. *Asian Social Science, 7*(12), 163–167.

Chtcheglov, I. (1953). Formulary for a new urbanism. Reprinted in K. Knabb (Ed.), *Situationist international anthology*. Berkeley, CA: Bureau of Public Secrets, 1995.

Currier, D. (2003). Feminist technological futures: Deleuze and body/technology assemblages. *Feminist Theory, 4*(3), 321–338.

Debord, G. (1955). Introduction to a critique of urban geography. Reprinted in K. Knabb (Ed.), *Situationist international anthology*. Berkeley, CA: Bureau of Public Secrets, 1995.

Deleuze, G., & Guattari, F. (1987). *A thousand plateaus: Capitalism and schizophrenia* (B. Massumi, Trans.). Minneapolis: University of Minnesota Press.

Dewsbury, J. D. (2011). The Deleuze–Guattarian assemblage: Plastic habits. *AREA, 43*(2), 148–153.

Dovey, K. (2012). Informal urbanism and complex adaptive assemblage. *International development planning review, 34*(4), 349–367.

Grosz, E. (1994). A thousand tiny sexes. In C. Boundas & D. Olkoswki (Eds.), *Gilles Deleuze and the theatre of philosophy*. New York: Routledge.

Grosz, E. (2002). Bodies-cities. In G. Bridge & S. Watson (Eds.), *The Blackwell city reader* (pp. 297–303). Malden, MA: Blackwell.

Human Rights Watch. (2012). "Beat him, take everything away" — abuses by China's Chengguan para-police. Retrieved from www.hrw.org/reports/2012/05/23/beat-him-take-everything-away

Jian, Y. (2013, November 26). Night market forces buses to change routes. *Shanghai Daily*. Retrieved from www.shanghaidaily.com/Metro/society/Night-market-forces-buses-to-change-routes/shdaily.shtml

Jones, P. (2012). Sensory indiscipline and affect: A study of commuter cycling. *Social & Cultural Geography, 13*(6), 645–658.

Khatib, A. (1958). Attempt at a psychogeographical description of Les Halles (P. Hammond, Trans.). *Internationale Situationniste #2*. Retrieved from www.cddc.vt.edu/sionline/si/leshalles.html

Lingis, A. (1994). *The community of those who have nothing in common*. Bloomington: Indiana University Press.

Loyalka, M. (2012). *Eating bitterness: Stories from the front lines of China's great urban migration*. Berkeley: University of California Press.

Marx, C. (2009). Conceptualising the potential of informal land markets to reduce urban poverty. *International Development Planning Review, 31*, 335–353.

Meng, X., Manning, C., Shi, L., & Effendi, T. N. (2010). *The great migration: Rural-urban migration in China and Indonesia*. Northampton, UK: Edward Elgar.

Neuwirth, R. (2011). *Stealth of nations: The global rise of the informal economy*. New York: Pantheon Books.

Pálsson, G. (1994). Enskilment at sea. *Man, 29*, 901–927.

Portes, A., Castells, M., & Benton, L. A. (1989). *The informal economy: Studies in advanced and less developed countries*. Baltimore, MD: Johns Hopkins University Press.

Roy, A. (2011). Slumdog cities: Rethinking subaltern urbanism. *International Journal of Urban and Regional Research, 35*(2), 223–238.

Shanghai Municipal Information Office & Shanghai Municipal Statistics Bureau (2011). *Shanghai basic facts 2011*. Shanghai: Shanghai Literature and Art Publishing Group. Retrieved from http://en.shio.gov.cn/facts.html

Tsai, K. S. (2002). *Back-alley banking: Private entrepreneurs in China*. Ithaca, NY: Cornell University Press.

Wise, J. M. (2005). Assemblages. In C. J. Stivale (Ed.), *Gilles Deleuze: Key concepts* (pp. 77–87). Montreal: McGill and Queen's University Press.

Zhang, L. (2001). *Strangers in the city: Reconfigurations of space, power, and social networks within China's floating population*. Stanford, CA: Stanford University Press.

Zhenqi, Y. (2013, December 17). Good riddance to night market. *Global Times*. Retrieved from www.globaltimes.cn/content/832632.shtml#.UsD_G8uwrtR

10 Scarcity and the Making of Bottled Water Markets in Chennai

Emily Potter

On the streets of Chennai, Tamil Nadu, plastic water bottles appear everywhere. Scattered through piles of rubbish, clustered together on street sellers' wagons, piled next to doorways, and transported around the city on a host of dedicated water-vending trucks, bottled water has become a ubiquitous part of Chennai's streetscape. This ubiquity is not unique to Chennai, as bottled water consumption has skyrocketed in India since the beginning of the 21st century, and across Asia more broadly. Over 5 billion litres of bottled water are consumed annually in India, with sales totalling US$250 million in 2010 (Indian Bottled Water Market, 2011). Concurrently, bottled water consumption throughout the more affluent West has declined, as this product has come under a growing challenge from activists, government authorities, and, increasingly, the broader public, who point to the wasteful, unnecessary, and faddish characteristics of a product once labelled "a triumph of marketing" (Howcroft, 2009).

The recent rise of bottled water consumption in India has been critically framed in terms of the broader "environmental insanity" of bottled water's global success (Milmo, 2006), and, more specifically, of "capitalist urbanization" (Gandy, 2008, p. 116), and the reach of marketization into all realms of life. As one candid commentator puts it: "the boom time for Indian bottled water continues . . . because the economics are sound, the bottom line is fat and the Indian government hardly cares for what happens to the nation's resources" (Pai, 2006). This view of the Indian government having 'absented' itself from a primary responsibility to provide water to its population informs much debate on the subject, and is primed by the idea that in civil society water is free-flowing and indiscriminately available to all. This view of water has made it a highly contested 'object' of market designs, with a long history of politicization in India stemming from the earliest efforts by colonizing forces to control the supply and consumption of its water resources. The meta-critical discourse of bottled water consumption in India is informed by this history and draws heavily on the notion of water rights, which are placed in tension with the logic of the market (Shiva, 2002). This discourse frequently implies that market forces are incapable of ensuring—and indeed are in contravention of—a 'human right to water,' and the market's ability to fulfil the basic needs of human beings, without prejudice, is at the heart of much pro-market and anti-market debate when it comes to the question of water delivery.

Certainly, in contexts such as Australia, where bottled water commonly sells for one to two thousand times the cost of tap water (Bottled Water Alliance, 2010), the consumption of bottled water can be seen as a convenient supplement to a public supply system that is, except in some remote communities, universally accessible and of high quality. In many other contexts, however, where public water supply is far from guaranteed, bottled water takes on a different function—one that sees it positioned uncomfortably within the daily regimes of water provision that water rights advocates ideally associate with the networked tap. As it is played out in Chennai, this scenario has seen an informal network of containerized water provision emerge and thrive. For some commentators, this is purely symptomatic, the result of failing, or failed, piped water systems (Narain, 2005). For others, it represents a continuation of systems of private water vending that have always operated in poor, non-networked communities (Kjellén & McGranahan, 2006).

There is something more to the participation of bottled water in Chennai's water landscape, however. Drawing upon Ananya Roy's (2011) understanding of 'informality,' this chapter will discuss the growing and diversified consumption of bottled water in Chennai to explore the ways in which bottled water actually participates in the making of its markets. This is a significant shift from the bottle of water as supplement, salve, or villain, to instead see it as an active agent in the enactment of a reality in which water is scarce and crucially, as a result, is provided by a host of nonstate actors. 'The street' comes into play here as a site in which urban informality, in Roy's sense, is practised. It is the milieu in which different realities intersect and are produced, at once a zone of commerce, consumption, and inequitably dispersed power, and also, literally, the geography in which bottled water is most visible, as it goes about generating its markets on the back of trucks, at filling stations, or clustered on street stalls. The concept of enactment drawn upon here is informed by Michel Callon's (1999) theorization of a market *as* enacted— that is, a market is not a pre-given force, rather, it is brought into being by a host of techniques, materials, and players, and is consequently contingent and open to transformation. It is also relational, and certainly more-than-human. As the consumption of bottled water in Chennai demonstrates, the very informality of this city's water markets is registered in this interaction of the authorized and unauthorized, the secure and the insecure, and, amongst this, the bottle and the tap, both performing water access and absence.

WATER ACCESS IN INDIA

The nonsustainable, nonrenewable, and polluting plastic culture is at war with civilisations based on soil and mud and the cultures of renewal and regeneration. Imagine a billion Indians abandoning the practice of water giving at *pijaos* and quenching their thirst from water in plastic bottles?

(Shiva, 2002, p. x)

Prominent water activist Vandana Shiva's assessment of the increasingly widespread consumption of bottled water in India frames the bottle as a wasteful luxury few Indians can afford, materially, culturally, or environmentally. This view is reiterated by other commentators and activists for whom bottled water (like Coca-Cola before it) has become an icon of Western lifestyles. It is also an indication of the economic and social inequalities that capitalist consumption reiterates and generates in a country with relatively poor public infrastructure. In India's populous cities, public water systems afford far from universal access, and this uneven coverage and often-unreliable supply have meant that the ability to meet the water needs of millions of Indians is a daily challenge.

This "hydrological dystopia," as water geographer Matthew Gandy describes it (2008, p. 116), is commonly signified by the urban slum; in particular, with its lack of basic services, water, and sanitation. Here, as Gandy (2004) argues elsewhere, a "brutal distinction" between "citizens" and "mere 'subjects' " (p. 368) is marked by inequitable access to networked infrastructure. This lack of sufficient public water has a history in the country's shift to industrialization under British administration. While the British brought a vision of civic water supply, the reorganization of economic and social life under colonization saw traditional, localized and sustainable water-harvesting practices lose favour and capacity.

As a result, India's history of postcolonial water management has favoured supply-focused solutions; in particular, large-scale projects that divert, dam, extract, process, and transport water around the country. Critically, a turn towards groundwater extraction and use (as a result, in part, of industrialized agricultural practices and water intensive cropping) has most affected potable water availability across the country, declining from a per capita availability of 5,177 cubic metres in 1950, to 1,820 cubic metres in 2001 (Shiva in Ray, 2008, p. xi).

The World Bank has been active in this, subsidising mechanized withdrawal systems since the 1970s, and providing credit for the installation of tube wells to enable large-scale irrigation (Shiva, 2002, p. 10; Briscoe, 2005, p. 22). Tube wells—long stainless steel pipes that draw water up from the ground via electric pump—are so efficient that groundwater can be withdrawn much faster than it can be replenished. More than 20 million tube wells have been installed throughout India (Barlow, 2007, p. 13). This has resulted in widely overdrawn aquifers. Depleted groundwater resources cannot recharge surface water, which leaves groundwater as the only source of viable potable water. By the mid 2000s India relied upon groundwater for 80 per cent of its drinking water (Briscoe, 2005, p. 23).

Bottled water's arrival on the Indian market is integrated with the story of groundwater exploitation. Branded bottled water first appeared as a luxury product in India, much as it did elsewhere around the world, and flourished in the conditions of the economic liberalization of the 1990s. At this time, bottled water was routinely pitched to consumers as a boutique beverage option and a lifestyle choice, linked in its marketing with glamorous Bollywood celebrities; in 2001, the newspaper *India Today* proclaimed bottled water to be "wet and sexy" (Renuka, 2001). By early 2000, however, a different pitch had entered the marketing of bottled water in India, and with it the industry really took off, reaching levels of 40–50 per cent growth per year by 2005 (Murthy, 2005). This pitch was framed around concerns over water scarcity.

The chairman of Bisleri, one of India's leading bottled water brands, explained his product's success in these terms: "There was a clear opportunity for building a market for bottled water. The quality of water available in the country was bad. It was similar to what Europe faced before World War 2. The quality of water in Europe was extremely poor, which created a bottled water industry there. In India, too, not only was water scarce, whatever was available was of bad quality" (Raturi, 2005). Bisleri's 2001 catchphrase "Play Safe" enforced these associations; while its competitor Kinley's 2002 slogan *"boond boond mein vishwa"* translates to "reliability in every drop" (Murthy, 2005). These claims to reliability reference not just the quality of the water itself, but also its availability. As industry group Bottled Water India (n.d.) succinctly states, "the water shortage around the world and particularly in third world countries has opened new avenues for bottled water" (n.p.).

The emergence of scarcity discourse in the marketing of Indian bottled water occurred on the back of the packaging revolution created by cheap and durable polyethylene terephthalate (PET) bottles. As a result, it pushed bottled water production away from the more expensive single-serve bottle and towards the 20-litre bulk container. Bisleri first ventured into 20-litre bulk water in 2001, and by 2005 this product segment constituted 60–70 per cent of their entire bottled water sales (Raturi, 2005). Now Bisleri, and other leading bottled water manufacturers, offer a 24-hour home delivery service: "Picture this," a plug for this service explains, "It's the rainy season. You come home late, tired, and discover that your building doesn't have electricity and water and you have run out of drinking water as well . . . Just pick up your phone and dial Bisleri" (Polaris Water Blog, 2010). While this kind of service operates at the top end of the market, bulk water delivery to homes and workplaces has become a widespread business in India, and the 20-litre bottle has extended the reach of the bottled water market to lower income groups (Raturi, 2005). Nevertheless, it is the relatively unmonitored and cheap allocation of licenses to extract and bottle water for sale that has most significantly fuelled India's bottled water boom (Lakshmi, 2011). This water comes almost exclusively from groundwater sources. India currently has approximately 1,400, with 270 based in Chennai alone (Mariappan, 2011).

BOTTLED WATER IN CHENNAI

Between 2010 and 2011 the sale of bottled water in Chennai rose from 4 million litres per day to 6 million litres per day; by then, an estimated one third of Chennai's population used bottled water to meet their daily water needs (Mariappan, 2011). Over half of this amount (approximately 3.5 million litres) is bulk water sold in 20-litre bottles. At the same time, Chennai's public water network offers one of the lowest levels of water availability per day on average in the country (Gopakumar, 2012, p. 59), earning this city of almost 4.7 million people the moniker of India's "water scarcity capital" (Desai, 2010). Until the 1970s, the city's public water supply system had drawn exclusively on surface water from its system of reservoirs; in the preceding 20 years, the per capita availability from these sources had dropped by 40 per cent, from 140 litres per capita in 1951, to 80 litres per capita by 1971 (Vaidyanathan & Saravanan, 2001, p. 4). The burgeoning city meant that many peri-urban and some urban areas were without piped supply altogether. In an effort to keep up with demand, Metro Water (an autonomous statutory body charged with the provision of water to the city's residents) began to sink bore wells fitted with public hand pumps around the city, and started to draw groundwater (including the purchase of agricultural water from farmers) from an increasing radius around the capital.

By 2005, over a quarter of a million residents were officially outside the area of Metro Water's networked service, while the agency's supply met only one-third of total daily demand (Narain, 2005, pp. 4–5). Commonly, water is available through the tap for only a few hours each day, and, frequently, on alternate days or even once per week. In 2010, the suburb of Anakaputhur received piped water to its 25,000 households once every 15 days (Desai, 2010). The illegal tapping of pipes by individuals with booster pumps further compromises what flow there is. Tap water commonly smells and tastes unsavoury, and is full of corrosive minerals that damage pipes and prevent residents from using their showers. In many households, tap water, when it is available, is used only for bathing and other domestic tasks (C. Kurian, personal communication, 8 June 2009).

Given this, a multiplicity of bottled water practices coalesce in and through the streets of Chennai. These include a range of water vending and delivery devices, across a spectrum of economic means. Different types of plastic are used for different bottled products, with the cheaper end of the range utilizing thinner, more porous material; 200 ml bottles, or 'cups' as they are called, are sold cheaply for a few rupees only in high traffic contexts such as railway and bus terminals. For India's poorest inhabitants, the cost of much bottled water is prohibitive. However, even for this demographic—often the most likely to be without networked supply—water vending in other forms plays a central role in meeting daily water needs, incorporating street-side local water kiosks and private taps where people can refill plastic bottles for a price (Kjellén & McGranahan, 2006, p. 16).

For middle-class consumers, commonly with some limited tap water access (for short periods each day, for example), multiple 20-litre bottles are usually delivered several times a week (C. Kurian, personal communication, 8 June 2009). A newspaper report in 2009 profiled T. Swaminathan, who services the Chennai neighbourhood of Koyambedu, as an example of bottled water entrepreneurship. Swaminathan delivers 20-litre bottles on the back of his motorbike to wealthy gated communities, using a missed call system on his mobile phone to communicate with his customers—the door numbers of the houses he services are stored on his phone; when a customer calls twice, he knows that they require water and will promptly deliver to their door (Varma, 2009, p. 2).

The 'convenience' of bottled water, as opposed to the unpredictable, irregular nature of tap water, has contributed to its swift integration into daily life. However, its strong associations with health and reliability also remain. In 2009, the residents of the Chennai suburb Virugambakkam reported that their water had turned black and smelt of sewerage; several children fell ill with typhoid and malaria (Sujatha, 2009, p. 3). A local doctor 'prescribed' bottled water to local families as a response. This is despite bottled water coming under scrutiny for its own health dangers. Bulk water is usually minimally treated with reverse osmosis techniques and ultraviolet (UV) radiation (Murthy, 2005), and thus problems with contamination are common. This has haunted India's bottled water market in general since tests undertaken in Delhi in 2003 revealed unacceptable traces of contaminants in all locally bottled brands. In Chennai, the repeated reuse of containers by bottlers is a factor in ongoing concerns with water quality in 20-litre bottles particularly. Residents have reported falling ill with throat and respiratory infections after consuming particular brands (Lakshmi, 2011). Given the proliferation of bottlers, however, it is easy enough to move on to another one. Most, but not all, bottled water plants are licensed, and the Bureau of Indian Standards (BIS) issued 100 licenses in 2010. In a two-month period in 2011 it was reported that 50 applications for new licenses alone had been received by the BIS (Lakshmi, 2005).

The normalization of bottled water has meant that even where other sources of water are freely available bottled water continues its strong presence. Rainwater harvesting activist S. Raghavan, who has worked to restore the communal wells in his apartment complex so that its residents no longer need rely on either bottled water or degraded tap water, argues that bottled water is an "addiction." Even with local wells full of water, he reports, his neighbours still regularly purchase bottled water (S. Raghavan, personal communication, 9 June 2009).

The resulting increased demand for bottled water has seen water tanker trucks become a common sight on Chennai's roads, drawing water from agricultural land further and further afield for both the public network and water bottlers. The practice is so lucrative that oil tankers have been arriving from other states to join the water sale and transportation industry. It is widely known that some local politicians participate in the containerized water business, including ownership of some of the private water tankers that are hired out to Metro Water (S. Raghavan, personal communication, 9 June 2009). Other stories of criminal involvement in the water tanker truck industry report that organized syndicates sometimes collude with local police to take control of the bulk water brought in by Metro Water, forcing communities to pay above standard rates (Deccan Chronicle, 2009). Meanwhile, for many farmers groundwater has become their most profitable 'crop.' The culture of water containerization has been promoted by these tankers, which normalize mobility and the transfer of water between source and consumer via unpiped methods.

At the extreme end of bottled water normalization and reliance are Chennai's 'exclusive' housing estates being built on abandoned agricultural land at the city's periphery, drained of groundwater, and unconnected to Metro Water's network (Gopakumar, 2012, p. 63). These estates are *designed* to be serviced by bottled water, rather than a public network of supply. This scenario of broken-down systems and populations with no reliable access to water seems an apocalyptic one, echoed in Shiva's (2002) predicted "water wars" to come, as capitalism infiltrates the allocation of scarce, vital resources around the globe. Inequitable access to water is a given in the current arrangements of Chennai's water markets. Those without the capacity to pay, or without access to the network of tankers and private providers, must contend with poor quality and unreliable bore water, and the consequences of these. Yet what emerges from a brief mapping of tap and bottled water consumption in Chennai is not a picture of totalized capitalist triumph, but rather one of informal and varied water practices—including the tap—that dynamically enact a market in bottled water. This market is continually being made, brought into being by the forces of both water's presence and absence that the bottle generates.

These bottled water practices are entangled with the tap rather than distinct from it, from the refilled plastic bottle to the groundwater sources that fill both bottles and Metro Water pipes. Different realities are generated by different water practices; but these do not conflict, for they all go into making urban informality. Thus, while the experience of bottled water consumption varies across economic groups—there is obviously a distinct difference between home delivery of a premium brand and filling your own bottle from a neighbour's tap—the capacity to pay for and access water in these ways, and circumvent an unreliable network, creates realities of water abundance alongside the realities of water scarcity that the bottled water industry continues to enact. The idea of simply picking up one's phone and dialling Bisleri, or sending a text to a local water vendor, creates a reality in which water is free-flowing, if not free of charge. At the same time, the ongoing extraction of groundwater within Chennai's environs means that bottled water is inextricably caught up in the making of water-scarce realities—realities that bottled water also discursively enacts as it positions itself to consumers as a reliable salve to the ills of a dysfunctional piped network and drought-stricken environment. This city, after all, is India's 'water scarcity capital' *and* the epicentre of the country's industries in water containerization, transport, and sale.

As earlier indicated, the understanding of informality employed here is drawn from Roy's (2011) theorization of "urban informality" in the context of subaltern studies (p. 232). For Roy, urban informality is not so much an identification or disposition (often celebrated as resistant or subversive), but rather is a mode of producing urban space—and, by extension, urban environments, with all their materiality. It is a set of practices that, according to Roy, encapsulates "the ever-shifting urban relationship between the legal and illegal, legitimate and illegitimate, authorized and unauthorized" (p. 233). Urban informality encapsulates the shady realities (in the sense of not being clearly defined) of how people, from the rich to the poor, sustain themselves, claim or enact social and political power, and negotiate the administrations of the state. These "fractal geometries of metropolitan habitation" challenge the "territorial imagination of cores and peripheries" in which the informal is allied to the economically and socially marginalized subject, and the formal to the elite (p. 233).

Roy's analysis speaks to the complex networks of Chennai's water economy, in which informal water practices are normative across the socioeconomic board and not just the purview of the poor and disenfranchised. Chennai's residents, from the suburbs to the slums, access and use a mix of state and nonstate, legal and illegal, water apparatuses and resources in all sorts of arrangements in which containerized water recurrently figures. For instance, the new, expensive suburban developments on land depleted of groundwater, and beyond the reach of current piped infrastructure, depend on tanker trucks and 20-litre containers for their entire water resources. This water—like all containerized water transported and sold in Chennai—will not necessarily be from a 'legitimate' source, given the activities of criminal syndicates and illegally sunk boreholes in the city's water economy. Water accessed and supplied by Metro Water, as part of their public supply service, may also come from these dubious sources. Meanwhile, poorer residents will access public water from a local tap, while supplementing this (when the tapped supply is unreliable) with privately purchased containerized water from one of the city's many water bottlers.

These various arrangements of private capital, the state, community, and water that manifest in Chennai at a time of announced water scarcity, indicate the truth of Bakker's (2010) assertion that "multiple supply sources" characterize water provision in the cities of the Global South (p. 442). The increasing integration of bottled water into daily water regimes has caused concern amongst water commentators, who draw a clear distinction between the interests of the market and the responsibilities of the state. However, the realities of water access and supply in developing countries, and in cities such as Chennai, suggest that a distinction between mass 'public' water and boutique 'private' water are far from clear-cut.

The framing of bottled water as a wasteful Western luxury fails to attend to a reality in which bottled water has gone beyond a status or convenience object. Instead it has become a normalized, and even crucial, means of water delivery for many people, both affluent and poor. Importantly, bottled water is not a passive participant in the story of Chennai's "water woes" (*The Hindu*, 2003). It is an active force in the midst of an emergent water reality, triggering and shaping new water arrangements and behaviours, and, more broadly, arrangements of urban inhabitation. A reality, as Annmarie Mol (2002) might say, "does not 'stand by itself' but emerges in the relations of practice . . . and there are many practices" (pp. 31–32). In the case of Chennai, water scarcity and water abundance are therefore enacted, along with the markets and bureaucracies that serve and respond to these mutable conditions, and the bottle is at work in bringing these realities into being on the streets and into the homes of the city.

ACKNOWLEDGEMENTS

This research was funded by an Australian Research Council Discovery Project, "From the Tap to the Bottle: The Social and Material Life of Bottled Water." The author would like to acknowledge her co-researchers, Gay Hawkins and Kane Race, for the many conversations that fed into the research that is presented here.

REFERENCES

Bakker, K. (2010). *Privatizing water: Governance failure and the world's urban water crisis*. Ithaca, NY: Cornell University Press.
Barlow, M. (2007). *Blue covenant: The global water crisis and the coming battle for the right to water*. Melbourne: Black.
Bottled Water India. (n.d.). Retrieved from http://bottledwaterindia.org/about-us-2/
Briscoe, J. (2005). *India's water economy: Bracing for a turbulent future*. Washington, DC: World Bank.
Callon, M. (1999). Actor-network theory—the market test. In J. Law & J. Hassard (Eds.), *Actor-network theory and after* (pp. 181–195). Oxford: Blackwell.
Deccan Chronicle. (2009, September 27). OMR residents face acute water shortage. Retrieved from www.deccanchronicle.com/content/tags/chennai/27/9/2009.html
Desai, D. (2010, March 8). The looming water shortage. *Financial Chronicle*. Retrieved from www.mydigitalfc.com/leisure-writing/looming-water-shortage-386
Gandy, M. (2004). Rethinking urban metabolism: Water, space and the modern city. *City*, 8(3), 363–379.

Gandy, M. (2008). Landscapes of disaster: Water, modernity, and urban fragmentation. *Environment and Planning A, 40*(1), 108–130.

Gopakumar, G. (2012). *Transforming urban water supplies in India: The role of reform and partnerships in globalization.* London: Routledge.

Go Tap, Bottled Water Alliance. (2010). Are consumers being ripped off by the bottled water industry? Retrieved from http://dosomething.net.au/media/767/100702%20mr%20bottled%20water%20pricing%20finalwebcopy%20.pdf

Howcroft, R. (2009, April 2). *Gruen transfer* [Television broadcast]. Australian Broadcasting Corporation.

Kjellén, M., & McGranahan, G. (2006). *Informal water vendors and the urban poor.* Human Settlements Discussion Paper Series. Institute for Environment and Development, London.

Lakshmi, K. (2005). Chennai: A city in deep waters. *The Hindu.* Retrieved from www.hindu.com/pp/2005/04/16/stories/2005041600040100.htm

Lakshmi, K. (2011, March 2). All that's packaged isn't always drinkable. *The Hindu.* Retrieved from www.thehindu.com/news/cities/Chennai/article1501694.ece

Mariappan, J. (2011, April 12). City sees jump in sale of water cans. *The Times of India.* Retrieved from http://articles.timesofindia.indiatimes.com/2011-04-12/chennai/29409770_1_drinking-water-water-supply-mld

Milmo, C. (2006, June 29). "Environmental insanity" to drink bottled water when it tastes good from the tap. *The Independent.* Retrieved from www.commondreams.org/headlines06/0629-01.htm

Mol, A. (2002). *The body multiple: Ontology in medical practice.* Durham, NC: Duke University Press.

Murthy, L. (2005). *Boond-boond mein paisa*: bottled water is big business. Retrieved from http://infochangeindia.org/agenda/the-politics-of-water/boond-boond-mein-paisa-bottled-water-is-big-business.html

Narain, B. L. (2005). Water scarcity in Chennai, India. Retrieved from www.cccsindia.org/ccsindia/interns2005

Pai, U. L. (2006, July 8). Water—India needs massive investments. Retrieved from www.investorideas.com/Iil/News/080706.asp

Polaris Water Blog. (2010, June 15). India: 24 hour water delivery—for those who can pay for it. Retrieved from http://polariswater.blogspot.com.au/2010/06/india-24-hour-water-delivery-for-those.html

PRNewswire. (2011, March 24). Indian bottled water market expecting rapid growth. Retrieved from www.prnewswire.com/news-releases/indian-bottled-water-market-expecting-rapid-growth-118578544.html

Raturi, P. (2005). And this is how Parle Bisleri began. Retrieved from www.rediff.com/money/2005/jun/10spec.htm

Ray, B. (2008). *Water: The looming crisis in India.* Lanham MD: Lexington Books.

Renuka, M. (2001). Liquid asset. *India Today.* Retrieved from www.india-today.com/itoday/20010514/business-water/shtml

Roy, A. (2011). Slumdog cities: Rethinking subaltern urbanism. *International Journal of Urban and Regional Research, 35*(2), 223–238.

Shiva, V. (2002). *Water wars: Privatisation, pollution and profit.* Cambridge, MA: South End Press.

Sujatha, R. (2009, June 12). Contaminated water causes diseases, allege residents. *The Hindu,* p. 3.

The Hindu (2003, June 5). Chennai's water woes. Retrieved from www.hindu.com/thehindu/mp/2003/06/05/stories/2003060500290100.htm

Vaidyanathan, A., & Saravanan, J. (2001). *Managing water in Chennai.* New Delhi: Centre for Science and Environment.

Varma, D. (2009, June 11). The changing profile of "missed call" makers. *The Hindu,* p. 2.

11 Street Vendors in Cairo
A Revolution Orientated Strategy
Nashaat H. Hussein

INTRODUCTION

Street vending is not new to Egyptian society, yet very few sociologists have attempted to understand the real etiological base of the practice, and its direct influence on the lives of the poor in Egypt. Most of the studies that incorporate the term 'street vendors' consider the issue from a predominantly economic perspective (Chowdhury, 2007; El Mahdi, 2002a; 2002b; Harati, 2013; Moktar & Wahba, 2002; Rizk, 1991). They overlook the social dynamics that often lead to the prevalence of the practice, and the fact that street vending has its grassroots in the social history of Egyptians. It represents an ordinary scene found in almost all Egyptian streets and public squares; Eastwood (2007) notes that "drawings in an Egyptian tomb exhibited at the Field Museum depict a market of four thousand years ago with vendors selling fruit, vegetables, and fish" (p. 567).

Most economists who have attempted to study the informal sector in Egypt argue that various socio-economic factors combine to help maintain informality, and, consequently, illegality related to the practice. Attia (2009) considers the positive impact of the informal economy as an engine for poverty reduction and development. He notes that informal enterprises in Egypt constitute 82 per cent of the total number of economic units and the informally employed—40 per cent of the total labour force. From his perspective, people's involvement in the informal sector is a better choice than being unemployed. Several interrelated socio-economic factors such as rural–urban migration, poverty, and lack of employment opportunities lead to the perpetuation of informality among many Egyptian entrepreneurs (Galal, 2004; Singerman, 1995).

Although there is no precise figure for the number of street vendors in Cairo, it has been reported that Cairo "has 600,000 street vendors" (Bayat, 2013, p. 82). In Cairo's streets, the classical dynamics by which the informal sector in Cairo operates depend on street vendors congregating in specific areas of the city where potential customers can be found, and selling alternative items at cheaper rates than equivalents sold in the formal sector (or in shops with formal licences to operate). For instance, "prepared food sold by street vendors from small carts is a very important part of urban dweller's diet" (Khouri-Dagher, 1996, p. 115) because food items are more affordable than those offered by licensed food-shops.

The long history of conflict with law enforcement agencies necessitates constant mobility of street vendors from one location in the city to another, displaying items to a mostly pedestrian customer base. This discourse of illegitimacy and informality has been challenged since the 2011 January Revolution. In the two years of lax security following the Revolution, the informal sector flourished in areas where people congregated *en masse*. Street vendors attended Tahrir and other public squares from the beginning of the protests, and adapted themselves to the nature of the events by selling the basic requirements of protestors during long hours of protest (like water, beverages, peanuts, pulp, T-shirts with revolutionary slogans, the Egyptian flag).

This chapter considers the strategy adopted by street vendors in Cairo in response to the protests following the January Revolution. Through in-depth interviews with a street vendor from Cairo, the chapter argues that although street vendors were not perceived positively before the Revolution, at least from the perspective of the government, the social and political contexts of the events not only affected their relationship with space and how they marketed their items, but also had a direct impact on how protestors perceived their presence and practices. Indeed, street vendors were not detached from events in the squares or from the protestors; rather they became part of the protest. The research concludes that the nature of the items they sold, particularly those related to the requirements of protestors, the locations they marked out in the squares, the nature of their mobility among protestors, and the protestors' interactions with them, all reveal a situation of co-habituation and coexistence with the events.

METHODS

The data here was collected through qualitative in-depth interviews with a street vendor living in Cairo. Tarek (a pseudonym) sells several items in the streets. He is 23 years old, married, and the father of one child. The research's aim was explained to him, and he agreed to be interviewed.

Tarek has been working as a street vendor for seven years. He did not finish his secondary education due to a lack of financial resources and because as he said, "he did not value education." Two of his friends in the residential area where he lives introduced him to street vending. They offered him a location (on a pavement); some items (fake eyeglasses) to sell in one of the streets off Ataba, another famous square near Tahrir in the heart of Cairo, and protection from other street vendors who might not accept his presence among them as a competitor. In time, he began to sell other items in different locations. Most of the commodities he sold were displayed on a wooden tray, which offered him a high degree of mobility in the case of attacks by municipal police. Bayat (2013) argues that "while the municipal police drive around to remove street vendors—in which case the vendors suddenly disappear—the vendors normally return to their work once the police are gone" (pp. 82–83).

Tarek believed that street vending is temporary work useful to save money to eventually buy or rent a shop. He admitted that most of his colleagues believe the same thing, and that many of them have succeeded in achieving their aim. The January Revolution meant a lot to him, because it was the first time he felt secure in the streets of the city owing to the absence of police attacks. He believed that most of the street vendors who attended the Revolution and the following mass congregations in the squares were attracted by the Revolution and its basic slogans: food, freedom, and social justice. He also stressed that most of the items sold at these sites supplied the needs and purposes of revolutionary-based protestors.

I carried out the interviews on the streets. A thematic interview guide was developed to cover the various themes related to Tarek's perception of street vending, the changing role of street vendors before, during and after the Revolution, and the way he perceives his future. The main data collection method used was semi-structured face-to-face in-depth interviews. The interviews took place between May and September in 2012. Illustrations in the chapter are included to provide descriptive evidence of the themes covered during the interviews, and to represent the attitudes revealed by the respondent. All illustrations are derived from the in-depth interviews. The interviews took place in colloquial Arabic, and I then translated them to English. The respondent was given a fictitious name as an ethical procedure to protect his identity.

STREET VENDING IN CAIRO'S STREETS: A SOCIAL SURVIVAL MECHANISM

In Egyptian culture, people usually make a distinction between 'legitimate' and 'illegitimate' work, especially when discussing work in the informal sector. Working casually in informal industrial units, in construction, in domestic service, or in selling items in street markets or in the streets, while viewed as being illegitimate by the government, is usually perceived as being legitimate or 'socially acceptable' by the public. From a social perspective, these types of activities are often treated as survival mechanisms or solutions for poor and needy families to earn a living and to avoid being involved in criminal activities.

In an interview with Tarek, he admitted that street vending is somewhat accepted socially because it represents a "justified solution to deal with poverty and lack of employment opportunities." Sympathy for street vendors is a contributing factor to their presence in Cairo's streets, and in the streets of other cities in Egypt:

> Our presence in the streets does not offend people. On the contrary, we have our own customers. People sympathize with our situation and understand that we are poor and we mean no harm to them. Our problem mainly comes from police attacks, and particularly from the owners of shops in areas where we exist. They believe that we compete with them by selling similar items at cheaper prices.

In Cairo, street vending is not typically perceived as a familial or a cultural practice. Additionally, selling items on a pedestrian level, especially in heavily populated districts, minimizes space between people and remote local markets. Food items, for example, can reach families in local districts through street vendors, which give people the opportunity to negotiate over prices—an advantage that may not be allowed in shopping malls or local markets. As Tarek explained, "Many street vendors sell vegetables and food items in the streets in poor and heavily populated areas. People usually depend on them instead of walking a long distance to local markets."

Being a street vendor in Cairo's streets is not typically an easy or affordable solution for the poor. It requires involvement in informal networks of micro-enterprises that guarantee street vendors protection, and a small sum of money to buy items directly from retailers to sell in the streets for a small profit. Using certain locations in the city to sell items also needs permission from the informal controllers of those areas. Territoriality is evident, and street vendors usually pay *ardia*, a daily tribute to the controllers to use certain pavements or certain locations in the city. In other words, street vending entails passing through long-term processes of mediation and appropriation with various informal actors in the streets. The daily dynamics of how street vendors operate in the city reflect a process of constant mobility and flexibility in dealing with street characteristics and various informal components.

STREET VENDORS AND REVOLUTIONARY PROTESTS

Revolutionary crowds in different squares in the city proved to be excellent business opportunities for street vendors, who discovered that no one would obstruct them doing business on the pavements of Cairo's main shopping streets. From the early days of the protests, street vendors congregated in Tahrir Square, where they created a flourishing local economy that provided for the hundreds of thousands of protestors who descended on the Square. For example, the protestors who congregated in Tahrir Square during the first days of the Revolution in 2011 mainly relied on street vendors for snacks and beverages, rather than the expensive cafes and restaurants of downtown Cairo.

Figure 11.1   A street vendor wearing paraphernalia related to the revolution sells food items in a protest.

(*Source:* Photo from the author's collection)

Most of the street vendors who assembled in the Square during the early days of the protests were those who used to sell their items in or around the Square and downtown area prior to the Revolution. In time, the location began to attract more street vendors from other locations in the city like Tarek, who said "I used to sell my items in a main street downtown. When we realized that people began to assemble in the Tahrir Square, we moved there."

Feelings of anxiety and threat created a mutual sense of interdependency and cohesion between the protestors and street vendors. As the protests progressed, new items began to appear in the squares other than food, water, and beverages. Supported by an informal industry responsive to events, street vendors began to sell flags with certain emblems (words, photos, colours), as well as masks, helmets, tents, and other items that matched the requirements of escalating protests.

Street vendors soon spilled onto the pavements and streets of the whole downtown area. An absence of formal social control encouraged thousands of people to exhibit and sell alternative items on the streets of Cairo. Successive assemblies of people in the hundreds of thousands (25 *en masse* gatherings of this scale were held between 2011 and 2013) in Cairo's main squares represented a significant opportunity for sales. Tarek outlined the differences between the types of items sold in the streets of Cairo before and after the January Revolution:

> Before the Revolution we used to sell several items to the people on the streets. There was no specific orientation to the types of the items we sold. You could find people selling food, vegetables, paper tissues, children's dolls, eye-glasses, etc. We used to suffer a lot from the attacks of municipal police forces on the one hand, and on the difficulties associated with convincing people of the quality of the items we used to sell on the other. During the protests, however, the items we sell mostly relate to the nature of the protests and the economic backgrounds of the protestors who mostly come from middle and lower-class families.

Recognizing that people's communication and interaction with each other usually leads to modifications in their ideas and behaviours, and

Figure 11.2   A family business where items related to the protests are sold.

(*Source:* Photo from the author's collection)

that their exposure to common threats may develop a conscious collective among them, the involvement of street vendors in the events, and their direct contact with protestors, seem to have affected their political orientations and organizational structure. In post-revolutionary Egypt political circumstances developed rapidly, and on September 26, 2012 street vendors in Cairo set up the Independent Street Vendors Syndicate. Their president claims that they already have 5,000 members, and are in the process of developing better solutions for the problems facing street vendors, not only in Cairo, but in other governorates of Egypt. Although the Street Vendors' Syndicate is still developing, its existence prompted the Syndicate of Commercial Professions, along with the Ministry of Finance and the Egyptian Trade Union Federation, to start counting street vendors nation-wide in an effort to consider the numerous vendors as beneficiaries of the new social insurance law. The Street Vendors' Syndicate currently attempts not only to officially register its members and unify the political voices of the vendors, but also to provide solutions to the problems they encounter while vending in Egypt's streets.

Street vendors have been attached to the January Revolution. Their sense of how space might be used changed dramatically, and they adopted a strategy in the squares of selling the requirements of protestors, as well as believing in the values promoted by protestors. This strategy, along with their emotional support of the revolution, had a direct impact on the way people perceived their existence, as Tarek stated:

> We became part of the Revolution. People in the past used to sympathize with us. Today, we began to feel that we support them. We are part of the society.

CONCLUSION

The goal of this study was to examine the impact of the Egyptian Revolution of 2011 on street vendors in Cairo, and the strategies they employed to respond to public protests and mass revolutionary congregations in the city. To understand the topic, I briefly focused on the social history of street vending in Cairo and people's perception of the activity as a social survival mechanism for poor and underprivileged families.

Findings indicate that the socio-historical development of the phenomenon of street vending in Cairo is based on two main factors: first, the social recognition of the phenomenon as a form of 'legitimate work' (if compared with other illegal means of earning a living); and second, people's sympathy with street vendors, based on an assumption that there are few other choices to survive given the increasing rate of poverty and unemployment in Egypt. Therefore, street vending in Cairo's streets has become an ordinary scene not typically resisted by the people, except when it comes into conflict with other formal economic activities or when street vendors begin to represent a threat to the flow of traffic or to public safety.

Findings also reveal that the eruption of the January Revolution had its direct impact on street vendors in the city, who found no socio-political resistance to their presence in areas where people congregated to protest (such as Tahrir Square), and who began to reorient their merchandise to suit the nature of the protests and needs of protestors. Despite initial differences in attitudes to the Revolution between street vendors and protestors, their existence together in squares at subsequent dates created a sense of commonality and a relationship among them. Feelings of interdependency and social cohesion developed between street vendors and protestors, and, in time, the entire strategy of street vendors, supported by a strong informal industry and trade, began to be directed to the revolutionary requirements of protestors. The Egyptian flag, T-shirts, masks, food, and beverages gradually invaded the squares used by street vendors, who began to accept the ideas raised by the revolutionary masses.

Although things will gradually return to normal, and street vendors may face the hardships they encountered before the Revolution, there is hope that they will enter the political arena and introduce new ideas and policies which allow them to have permanent shops located in specific areas in the city and to sell their merchandise, earn a living, and gain political support by forming their own professional syndicate.

REFERENCES

Attia, S. (2009). *The informal economy as an engine for poverty reduction and development in Egypt.* Munich Personal RePEc Archive. Retrieved from http://mpra. ub.uni-muenchen.de/13034/

Bayat, A. (2013). *Life as politics: How ordinary people change the Middle East.* Stanford, CA: Stanford University Press.

Chowdhury, S. (2007). *Everyday economic practices: The "hidden transcripts" of Egyptian voices.* London: Routledge.

Eastwood, C. (2007). Street vendors. In A. F. Smith (Ed.), *The Oxford companion to American food and drink* (p. 567). Oxford: Oxford University Press.

El Mahdi, A. (2002a). The labor absorption capacity of the informal sector in Egypt. In R. Assaad (Ed.), *The labor market in a reforming economy: Egypt in the 1990s* (pp. 99–130). Cairo: American University in Cairo Press.

El Mahdi, A. (2002b). *Towards decent work in the informal sector: The case of Egypt.* Geneva: Geneva International Labour Office. Retrieved from www.ilo org/wcmsp5/groups/public/@ed_emp/documents/publication/wcms_122058.pdf

Galal, A. (2004). *The winners and losers from the merging the informal economy in Egypt.* Washington, DC: Center for International Private Enterprise.

Harati, R. (2013). *Heterogeneity in the Egyptian informal labor market: Choice or obligation?* Working Paper, Centre d'Economie de la Sorbonne. Paris: Sorbonne University.

Khouri-Dagher, N. (1996). The state, urban households, and management of daily life: Food and social order in Cairo. In D. Singerman & H. Hoodfar (Eds.), *Development, change, and gender in Cairo: A view from the household*. Bloomington: Indiana University Press.

Moktar, M., & Wahba, J. (2002). Informalisation of labor in Egypt. In R. Assaad (Ed.), *The labor market in a reforming economy: Egypt in the 1990s* (pp. 131–158). Cairo: American University in Cairo Press.

Rizk, S. (1991). The structure and operation of the informal sector in Egypt. In H. Handoussa & G. Potter (Eds.), *Employment and structural adjustment: Egypt in the 1990s*. Cairo: American University in Cairo Press.

Singerman, D. (1995). *Avenues of participation: Family, politics and networks in urban quarters of Cairo*. Princeton, NJ: Princeton University Press.

12 Mapping Kuala Lumpur's Urban Night Markets at Shifting Scales

Khalilah Zakariya

In *The Bazaar: Markets and Merchants of the Islamic World*, Weiss (1998) wrote elaborately of the richness of the bazaar from four key perspectives: how the bazaar relates to the philosophy of life; its reflection of society; the buildings and layout; and its role as "the elixir of life." His in-depth narratives and descriptions of the bazaar connect the bazaar as a physical place with historical and cultural context, meanings, and experiences to reveal the bazaar's complexity and richness. Weiss wrote in the foreword of his book:

> The bazaar is much more than just a picturesque maze of workshops and shops in which tourists pick up souvenirs and get lost. It is a city within a city, with its own economy and way of life and a spiritual background from which western society has a great deal to learn.
>
> (p. 7)

The richness and complexity of a night market can be appreciated from a similar multilayered perspective. This chapter investigates night markets in Kuala Lumpur, Malaysia, and maps a night market operating at multiple scales: the city, street, and the stall. The scales are interrelated. Operations and changes that occur at the street or the stall scale will affect how the night markets operate at the larger scale, and vice versa. The aim of this chapter is to encourage designers, planners, market stakeholders, and even visitors to reflect upon how the night markets work in urban Malaysia, informing decisions made in regards to them.

THE NIGHT MARKET AS AN INFORMAL URBAN SPACE

The night market is an informal urban event commonly found in most Malaysian cities, towns, and neighbourhoods. Night markets, which usually take place from late afternoon until late evening, started operating in cities in

Malaysia in the 1970s following the implementation of the New Economic Policy (1971–90). City councils fostered the night market as a platform for small entrepreneurs and farmers to sell their products as a source of income, and as a result urban night market activities have been assimilated into the Malaysian culture.

In Malaysia, people are attracted to night markets to experience a part of the local culture. People can find local delicacies alongside localized 'global' dishes such as burgers, ice-blended drinks, and gas-oven pizza. People 'bump' into each other. Night markets are social events. The market organizer positions food and drink stalls among clothes, accessories, and toys stalls. The effect is a variety of social and cultural experiences for visitors and vendors, as well as buying and selling opportunities.

While I acknowledge the *pasar malam* image of the night markets, I contest the hegemony of this interpretation. *Pasar malam* means night market in Malaysia, yet it also connotes something very informal and casual. The expression conjures up representations of a chaotic, crowded, and lively night market atmosphere. I want to consider the formalities of such markets and how the formal and informal work together. This line of thought allow us to rethink how night markets in Kuala Lumpur operate as part of the city, as a way of knowing city life.

THE NEED TO REDISCOVER

The cultural geographer Doreen Massey proposes that we need to understand how places change (Massey, 1991; 2005). Massey explains that places are not static, but rather go through processes and accumulate meaning accordingly. Massey (1991) writes:

> What gives a place its specificity is not some long internalised history but the fact that it is constructed out of a particular constellation of social relations, meeting and weaving together at a particular locus.
>
> (p. 28)

Massey's work urges us to rethink how places like the night markets and other informal street markets work. They need to be valued for their dynamic qualities that are a result of their progress over time, but that also drive that progress.

Norberg-Schulz (1996) explains that the character of place is represented as "a general comprehensive atmosphere" and "the concrete form and substance of the space-defining elements" (p. 419). He relates this character of place to the experience of a person who visits a foreign city and notices particular characteristics of the city. This, in turn, shapes the person's experience. This sense that a place evokes in a person constantly refers back to certain unique characteristics that are different from other places. While designers and planners often search for a sense of place, Massey (1991) argues that such a "desire for fixity and for security of identity in the middle of movement and change" is problematic (p. 26). It would require places to have clear and rigid 'boundaries,' and the 'inside' and 'outside' of places would have to be distinguishable. Places would have to revert to having a single identity. Yet, informal urban street markets repeatedly demonstrate that there is always movement, change, dynamism, and porous boundaries.

METHODS: OBSERVING AND MAPPING NIGHT MARKETS

The investigation began by mapping 95 night market locations in Kuala Lumpur. Mapping is a method that records and reveals information to make invisible processes and characteristics visible. While maps traditionally function to record information or routes, maps also have qualities that reveal relationships between places and people (Harmon, 2004; Harmon & Clemans, 2009). When we choose to record specific things, we can 'see' and analyze how one thing relates to another. Black (2009) suggests that "mapping constructs a way of seeing" (p. 27). For example, choosing 'what' to map is critical to what and how a map might reveal.

Many artists and designers employ different modes of exploring and following certain objects, people or routes as methods to map the city (Calle & Auster, 1999; Careri, 2002; Hoffman & Irmas, 1989; Sadler, 1998; Wiley, 2010). From knowing the different ways people utilize, appropriate, or even ignore places, designers can then decide how and where to act. From the city to the street, the investigation followed a night market vendor for five days to his five different market sites. Through tracking the journey of a night market vendor, infrastructures and systems of the night markets became visible. The mapping process moved from the scale of the city to that of the street and the stall.

This study applied techniques of observation to understand how the vendor appropriates his temporary stall space. Observations were documented and photographed, and then mapped as a way of making connections between the night market, the city, and the users (Hood, 1997; Zeisel, 1984). Places like the night market are informal, temporary, and mobile. Visitors and vendors experience the senses of the night market at a human scale. When the market changes location, another spatial experience is created. Therefore, this study employed a more personal engagement with the night market sites in order to observe the market operation closely and broadly on shifting scales.

MAPPING NIGHT MARKETS IN KUALA LUMPUR

According to the 2009 report by the Petty Traders Development and Management Department of Kuala Lumpur City Hall, there are 95 night market locations operating throughout the city every week (Figure 11.1). This means that there is a night market operating somewhere in Kuala Lumpur every night. The total number of registered and licensed stalls is 31,652. This was a significant finding in this study because it demonstrates the extent to which night markets are a part of the Malaysian culture. These night markets operate legally under the license of operation given by the council. Different associations manage the different night markets.

Two of the longest running night markets are located in Lorong Tuanku Abdul Rahman (Lorong TAR) and Taman Connaught. The former is located in Kuala Lumpur, and the latter in a town located 12 kilometres from Kuala Lumpur. Both of these night markets are almost one kilometre long.

THE CASE STUDY: LORONG TAR NIGHT MARKET, KUALA LUMPUR

The Lorong TAR night market is a weekly night market that only operates on Saturdays. This night market is famous because it is the only one located in the city centre. In 2009, it hosted 552 market stalls. The Lorong TAR night market is a periodic and temporary market, where vendors only occupy the market site during the designated time. In this case, official trading hours allocated for the night market are from 6:00 p.m. until 12 midnight every Saturday. However, vendors already occupy the street and start setting up their stalls as early as 1:00 p.m.

VENDOR J AND THE NIGHT MARKETS

Vendor J is a forty-something man who sells at seven different markets in one week. He changes location every day. He has two stalls; at one he sells cakes, and at the other he sells kaftans. On days when he opens up both stalls, he will hire another vendor to operate one of the stalls. The site conditions of the night markets where Vendor J works differ according to locations. The Lorong TAR night market and the Jalan Kuching night market operate on a linear street. On the other hand, the Keramat night market operates in an open parking lot. In another location at the Shah Alam weekend daytime bazaar, Vendor J's stall is among the many market stalls operating at a large open-air parking lot of a stadium.

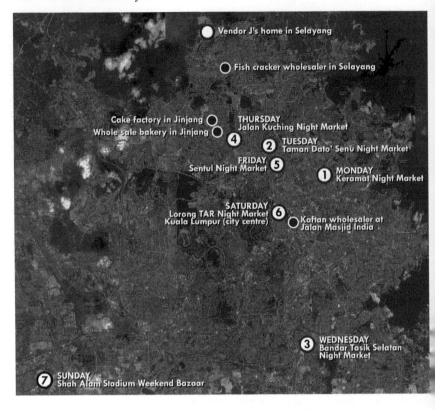

Figure 12.1   Vendor J's weekly market locations and supplier of goods.
(*Source:* Map data © 2014 Google, Sanborn)

MAPPING INFRASTRUCTURES AT THE CITY SCALE

Site mapping alongside Vendor J draws our attention to how night markets have 'hard' and 'soft' infrastructures. Hard infrastructure refers to components that physically support and enable the market operation. This includes location, surrounding context, the market space, and facilities that vendors and visitors find useful such as parking spaces, public toilets, electricity supply, and prayer rooms for Muslims. Adjacent neighbourhood contexts, such as commercial areas, residential areas, or tourist sites, are also vital components in the market system, as they act as the points of distribution for potential market visitors. In a way, the night market 'borrows' the functions of these more permanent and formal facilities.

The markets where Vendor J operates range from the neighbourhood night market to the city night market. Both types of markets have similar infrastructures. Neighbourhood night markets do not usually require public toilets or prayer rooms because they are located near to the residential areas, where there is usually a mosque or community prayer hall, as well as public toilets. At city night markets, visitors and vendors can go to the nearest mosque or the public prayer rooms provided at shopping centres, while at the same time using the public toilet facilities. Five of the markets where Vendor J operates are located near a mosque or a community prayer hall.

The location of Lorong TAR night market is strategic because support facilities surround the market. The night market is within the vicinity of several commercial places in Kuala Lumpur: shopping complexes, fabric and clothing retail shops, a wholesale supermarket, and tourist attractions such as the Masjid Jamek (Jamek Mosque) and the heritage buildings along the main road. The Lorong TAR night market not only shares the hard infrastructure from its surrounding neighbourhood, but also 'borrows' the familiarity and popularity of this heritage city quarter. This location is also convenient for local and foreign visitors because of its proximity to hotels and public transportation.

Hard infrastructure is tied to soft infrastructure. For a city, a component like a policy or a guideline is an example of soft infrastructure. In *Invisible Infrastructures*, Ware writes that policy "shapes the physical form of cities and communities" (2004, p. 122). The acceptance of a certain culture in a community is also a part of soft infrastructure, as it is an influential condition that encourages people to conduct certain activities. The familiar image of a place and the attachment that people have towards a place are also instances of soft infrastructure. They attach values to the hard infrastructure of the city.

It is clear that there are agreements and degrees of acceptance that enable night markets to operate in Kuala Lumpur at the city scale. In an interview with members of the Vendor Association of Lorong TAR Night Market, the president of the organization informed me that the planning and operations of the night market involve collaboration between their association, the city council, the vendors, and other support services. This process reveals that the night market, while including informality, also involves organization. Processes and procedures to plan a market either goes from a top-down approach, where the city council requests a night market to operate, or from a bottom-up approach, where the vendor association proposes a night market.

For Lorong TAR night market, the collaboration between the three main parties demonstrates the involvement of individuals and organizing bodies at different levels. For instance, while individual vendors must apply for their vending licenses from the city council, they pay their stall fees to the vendor association. The city council monitors the licensing and regulations. The vendor association manages and monitors the operations of the market on the site. Other services come into this process to support the night market, such as the neighbourhood watch, community association, garbage collection service, and electricity supplier.

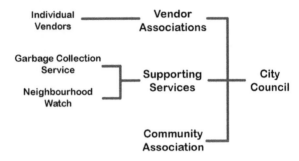

Figure 12.2   Components of the night market soft infrastructure.

In the case where the vendor association proposes a night market, it is required to submit a proposal to the city council. The council evaluates the proposal based on demand from the community, the flow of traffic that might change during the market operation, and the possible effects of the market on the surrounding business premises and neighbourhood. The relevant departments in the city council assess the proposed operations of the night market prior to approval or disapproval. Upon approval, the city council designates a site for the night market and a signboard is erected to inform the public about the market's operation. The council then demarcates the space for each stall based on the number of vendors participating in the night market.

The tracking and mapping activities of the night markets were conducted for five days around Kuala Lumpur, with one market outside of the city in Shah Alam. The network between Vendor J and his night market stalls is more robust than stalls and products. His products are a part of the larger wholesale product system and the formal industry.

Arriving at Lorong TAR, Vendor J sets up his cake stall first and then the kaftan stall. The step-by-step installation process is as follows:

Step 1: Park van temporarily near the stall lot

Step 2: Unload the tents from the van

Step 3: Open the tent that is approximately 3.6 meters by 2.4 meters (requires two people)

Step 4: Set up foldable table footing and unload the packs of fish crackers

Step 5: Arrange the plywood table-tops onto the table footing

Step 6: Unload the stacks of crates filled with cakes (requires two people)

Step 7: Unload two large containers filled with plastic bags, coins, light bulbs and cords, and store them under the tables.

Step 8: Cover the table with table mats

Step 9: Arrange the cakes according to the types—banana cakes on the far end, followed by cheese cakes, marble cakes and others. His assistant arranges the cakes in rows and stacks the similar types of cakes on top each other to maximize the space.

Step 10: While arranging, some customers are already visiting the stall to purchase the cakes

Step 11: Convert the unused crates into a stool and stores the remaining cakes under the tables

Step 12: Plug-in lighting to the nearest plug points

Vendor J's installation process uncovers further hard and soft infrastructures at the scale of the stalls. Foldable, mobile, and modular components are necessary for Vendor J because his kit of parts and products needs to fit inside his van. It was interesting to observe how Vendor J arranges the cakes that he bought from the wholesale bakery and cake factory. He stacks the

Figure 12.3 Vendor J's cake stall installation process.

cakes according to type on top of each other. The flat surfaces of the packaging made it easier for the cakes to be stacked so that Vendor J does not have to add an additional display rack. At the kaftan stall, he hangs the hangers on the stall frames. Here, the stall frames becomes an armature for vending. His appropriation techniques demonstrate that the kit of parts for vending at a night market needs to be adaptable to the products, and vice versa. In designing props and equipment for market vendors, modularity and practicality of installation must be considered.

While Vendor J sets up the cake stall, his assistants arrange the products at the kaftan stall. The kaftan stall involves the use of three tents: a large tent (3.6 × 2.4 metres) and two small tents (1.8 × 1.8 metres). The assistant vendors hang the kaftans around the three sides of the stalls on clothing hangers that attach to the tent frames. The assistant vendors then partially wrap a clear plastic around the stalls to create a thin wall between the kaftan stall and the adjacent stalls. This plastic cover acts as a separator from the fried chicken stall next to it while at the same time still allowing the colourful kaftans to be seen.

At the scale of the stall, micro infrastructures become apparent. Taking Vendor J's cake stall as an example, the hard and soft infrastructures for this one stall involve several components. The operation of a stall at Lorong TAR night market includes different fees. Vendors pay the space rental and purchase of standardized tents to the vendor association. They pay the fees for electricity and garbage collection services to a different organization.

The stall lot at this night market costs MYR52 (US$15) for one year, rented from the vendor association. Each vendor's license costs MYR26 (US$8) for one year paid to the city council. The vendor then purchases a standardized tent with the colour theme and logo of the city council and the vendor association for MYR235 (US$71) per stall. However, the vendor can use additional nonstandardized tents provided that they are extensions of the standardized stalls. This is a method for the association to monitor and regulate registered vendors. For the use of electricity, the city council

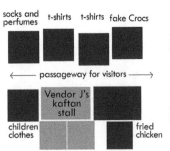 

Figure 12.4 Vendor J's kaftan stall.

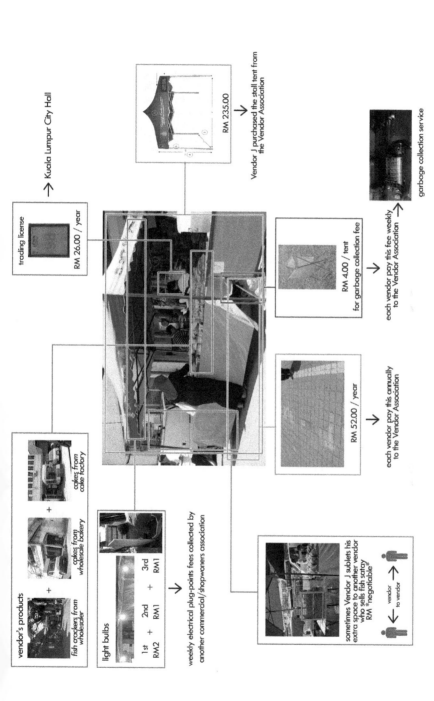

→ Kuala Lumpur City Hall

trading license

RM 26.00 / year

Vendor J purchased the stall tent from the Vendor Association

RM 235.00

garbage collection service

RM 4.00 / tent
for garbage collection fee

each vendor pay this fee weekly to the Vendor Association

RM 52.00 / year

each vendor pay this annually to the Vendor Association

vendor's products

fish crackers from wholesaler

+

cakes from wholesale bakery

+

cakes from cake factory

light bulbs

1st RM2 + 2nd RM1 + 3rd RM1

weekly electrical plug-points fees collected by another commercial/shopwoners association

sometimes Vendor J sublets his extra space to another vendor who sells fish satay
RM *negotiable*

vendor → to vendor

Figure 12.5 Hard and soft infrastructures for Vendor J's night market stall.

provides plug points with a fee of MYR2 (US$0.60) for the first light bulb, and MYR1 (US$0.30) for each subsequent light bulb. A representative from another association collects the fee weekly at the night market. As with the garbage collection service, vendors pay MYR4 (US$1.20) per stall. According to the vendor association, certain markets include the fees for garbage collection and electricity into the fees for the stall rentals. This depends on how the vendor associations manage their night markets. Additionally, there are vendors who sublet part of their stall space to other casual vendors. This becomes a subsystem, the vendor making a 'gentlemen's agreement' with the casual vendor.

LESSONS FROM THE NIGHT MARKET

At a glance, the night market may appear to be *ad hoc* and informal but, in fact, it works *with* a formal and an organized system. The system is a unique one, as it requires the framework to be flexible enough to suit the evolving and dynamic nature of the night market. Mapping the hard and soft infrastructures of the night market has been key to revealing this process, and exposes how the night market fits into the larger system of the city through operations at the scale of the city, the street, and the stall. The infrastructures that they need to operate also demonstrate that the system of a night market differs from other public spaces or commercial developments, due to its informality and temporality. With these observations in mind, urban planners, designers, and market organizers can better identify where and how to act.

Vendor J's journey map reveals a network of systems that exist as part of his night market operation. The network includes points of distribution where he sourced his products before selling them at the night markets. The map reveals the connections between Lorong TAR night market and systems of the city. Each night market vendor's operation contributes to the Lorong TAR night market at large. Vendor J purchases his night market products from manufacturers and suppliers and then sells to the customers at the night market. The products become part of the elements that give the night market its informality and richness, yet they may come from wholesalers in the formal sector. The different range of product choice and the variety of ways that the vendors display them create the colourful and rich image of the night market. This is part of the essence that makes the night market a vibrant event in the city.

REFERENCES

Black, R. (2009). *Site knowledge: In dynamic contexts*. Doctoral dissertation, RMIT University, Melbourne.
Calle, S., & Auster, P. (1999). *Double game*. London: Violette.

Careri, F. (2002). *Walkscapes*. Barcelona: Gustavo Gili.

Harmon, K. (2004). *You are here: Personal geographies and other maps of the imagination*. New York: Princeton Architectural Press.

Harmon, K., & Clemans, G. (2009). *The map as art: Contemporary artists explore cartography*. New York: Princeton Architectural Press.

Hoffman, F., & Irmas, D. (1989). *Sophie Calle: A survey*. California: Fred Hoffman Gallery.

Hood, W. (1997). *Urban diaries*. Washington, DC: Spacemaker Press.

Massey, D. B. (1991, June). A global sense of place. *Marxism Today*, pp. 24–29.

Massey, D. B. (2005). *For space*. London: Sage.

Norberg-Schulz, C. (1996). The phenomenon of place. In K. Nesbitt (Ed.), *Theorizing a new agenda for architecture: An anthology of architectural theory 1965–1995*. New York: Princeton Architectural Press.

Sadler, S. (1998). *The situationist city*. Cambridge, MA: MIT Press.

Ware, S. A. (2004). Invisible infrastructures. In J. Raxworthy & J. Blood (Eds.), *The MESH book: Landscape/infrastructure*. Melbourne: RMIT.

Weiss, W. M. (1998). *The bazaar: Markets and merchants of the Islamic world*. London: Thames and Hudson.

Wiley, D. (2010). A walk about Rome: Tactics for mapping the urban periphery. *Architectural Theory Review, 15*(1), 9–29.

Zakariya, K. (2012). Hard and soft infrastructures of temporary markets. In I. Zen, J. Othman, M. Mansor, & N. Z. Harun (Eds.), *Nurturing nature for man*. Gombak: International Islamic University Malaysia.

Zeisel, J. (1984). *Inquiry by design: Tools for environment behavior research*. New York: Cambridge University Press.

13 Territoriality in Urban Space

The Case of a Periodic Marketplace in Bangalore

Kiran Keswani and Suresh Bhagavatula

INTRODUCTION

'Territoriality' is defined as behaviour by which an organism lays claim to an area, creates a tangible or intangible boundary, and defends it against members of its own species (Hall, 1969). In this study, we try to understand how people generate such boundaries within urban spaces. In the literature on urban space, territoriality is understood as a spatial strategy to affect, influence, or control resources and people by appropriating area around them. It is a key geographical component in understanding how society and space are interconnected (Sacks, 1986). In our research, the context for understanding territoriality in urban space is the annual Peanut Fair in Bangalore, India. It is both a periodic marketplace and a cultural festival that is both informal and formal.

For those working within the informal sector, access to capital is limited, resources are not plentiful, and optimum use must be made of whatever is available. The informal worker therefore informally 'borrows' from the formalized public spaces of the city. No price, or a small price, for space can reduce or eliminate a prohibitive overhead related to rent/ownership of a space to sell from. There is an informal negotiation of space and formal arrangements during the annual peanut fair by vendors. They perform a balancing act to ensure a satisfactory environment for their customers (whose footpath they borrow), the formal shop owners (whose visibility they infringe upon), the municipal officials (whose planning regulations are not adhered to), and the police (whose law and order situation is made more complex through such informal activity). While the fair and market are officially sanctioned—spatially, socially, and economically—in the governmentality of the city the informality of how territory and services are appropriated is woven through this formality.

As people participate in the buying and selling of goods within the informal sector, they exercise control over the urban space in which they are situated. For vendors, this control is the ability to appropriate some part of the urban spatial environment to support their economic activity, and sometimes their social needs. It is this control exercised by street vendors that as understood in this chapter is seen as territoriality in the urban space.

The emergence and persistence of a periodic market in a region is often understood through economic location theory (Stine, 1962). This theory is concerned with the geographic location of economic activity where production and transportation costs are the primary influencing factors for a given location (Weber, 1929). However, periodic marketplaces are also influenced by social and cultural factors, particularly in their origin, their periodicity, and their continuity (Bromley, 1975).

This chapter discusses two kinds of territoriality perceived in an urban space. First, it looks at the utilization of the urban space for a periodic marketplace and how the fair marks its territory in the city. Second, it looks at how vendors informally mark and defend their own selling space or territory (Altman, 1975) within this urban space. It finds that collective memory, seen as a social and geographic construction of the past, plays an important role in this process (Rose-Redwood, Alderman, & Azaryahu, 2008). It highlights how this collective memory is an outcome of the social capital of street vendors informally accrued over generations and has become a key factor in the marking of territory. We argue that the social capital within this group is different from other forms of social capital. So, we introduce the term 'situational social capital' to capture this difference and subsequently describe its characteristics.

RESEARCH CONTEXT

The Peanut Fair at Basavanagudi in the city of Bangalore was chosen as the research context because it exhibits a process of territorialization developed through participation by many generations of farmers and traders from the surrounding villages. In the 12th century, Bangalore was itself a village, but is today the fifth largest metropolis in India, with a population of about 7 million (Sudhira, 2007). The Peanut Fair is a phenomenon that has taken place every year for the last 500 years. Vendors situate themselves on the street in spaces to which they have no formal rights. The Peanut Fair, colloquially called the *Kadalekai Parishe*, takes place during November and December each year. The fair began as a coming together of peanut growers to offer their first crop to the Bull Temple. Today, it is not just religious belief but also a business opportunity that attracts more than 200 peanut growers and traders to Bangalore.

In 1898, when a plague hit Bangalore, Basavanagudi became the location of the plague camp. Soon after, it was decided to plan a new layout on 440 acres of land here, which was then promoted as a model hygienic suburb (Figure 13.1). The Basavanagudi extension followed a rectangular design with boundary roads running north–south and east–west and the internal roads running parallel to them (Nair, 2005). This development circumvented the Bull Temple and was developed in such a way that the Bull Temple Road became the edge of the plan on its west side.

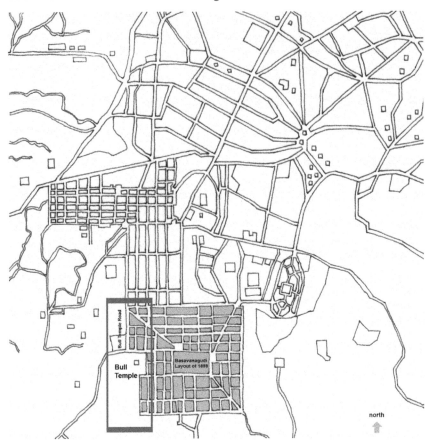

Figure 13.1   The new development of Basavanagudi extension circumvents the Bull Temple.

(*Source:* Based on the map of Bangalore from Murray's 1924 *Handbook*)

Today, the entire stretch of Bull Temple Road is pedestrianized for two days during the Peanut Fair. Vehicular traffic is rerouted and security arrangements are made to accommodate how the fair attracts visitors from all over the city and from neighbouring villages and towns. The Peanut Fair continues to be relevant for the city and presents an opportunity to understand the interconnectedness of social, economic, cultural, and geo-graphical relations within an urban space in a contemporary setting, as well as how informal arrangements reshape, produce, and rearticulate the formal arrangements of the city.

The Peanut Fair's location (the marking of territory) was determined by the religious need of peanut growers to offer their harvest to the temple. The periodicity is determined by a community decision to offer the first crop of the year at a time which is considered auspicious, on the last Monday of the Karthika month of the Hindu calendar. Its continuity is underscored by the need to keep the deity happy so as to ensure a good crop in the subsequent years. The periodicity of the fair and its location have also grown out of the need for a mutually convenient time and place for people coming from different villages. In the history of trading, the time and place for exchange of goods can become standardized and is often made to coincide with a day when people are likely to congregate for a social or religious activity (Bromley, 1975).

The offering of peanuts is made to the deity not only by the peanut growers but also by the residents of the neighbourhood. The residents play an essential role in the process of facilitation by being tolerant of the inconveniences caused by the changes in street usage during the fair. In the last few years, the preparation time for the fair has expanded, because the allocation of space works primarily on informal arrangements, sometimes on a first come, first served basis. As a result, access to the temple, to private residences, and to educational institutions along this road is affected for almost a week. The residents and visitors neither complain nor expect the situation to be any different.

During the field study, one of the questions asked of the local residents was: "The Kadalekai Parishe occupies the Bull Temple Road completely for two days. Do you think it should continue in this manner?" One resident responded: "Definitely it should continue in the same fashion, it's a part of our childhood." The residents would like to see the Peanut Fair continue because it originated centuries ago and has been there longer than they have. Arefi and Meyers (2003) explain that in India the perceptions of residents about public space is a reflection of tradition, cultural values, and practices of different social groups in the city.

For most of its history the fair was organized informally year after year by the growers themselves. However, it is now supported by two government institutions—the Department of Muzrai (Religious Endowment) and the Bruhat Bangalore Mahanagara Palike (BBMP) or the Municipal Corporation. The informal has interleaved with the formalities of a modernizing city. The local politicians direct the police force to cooperate during this time. There are 100 traffic police deputed for these two days to ensure good traffic management in the streets surrounding Bull Temple Road (from where traffic is diverted away) and 200 police officers to control the crowds within the pedestrianized zone. The security arrangements include a closed circuit television system to avoid thefts and physical barricades to ensure the separation of vehicular and pedestrian traffic. The hospitals nearby send out an ambulance in the case of an emergency and the Fire Department is on alert. This temporal pedestrianization of a high traffic zone constitutes a both a deterritorialization and a territorialization of the urban space. Such spaces are never one or the other but a negotiation somewhere in-between.

The peanut vendors occupy space on the sidewalk or footpath outside the formal shops. For the two days of the fair the shops lose visibility and lose some of their business. However, this is acceptable to the formal shop owners because the Peanut Fair is a religio-cultural phenomenon. Tolerance is their contribution towards its success and its continuity. Here again, there is a deterritorialization of formal arrangements that results from an informal social relation between the vendor and owners.

HOW VENDORS MARK THEIR TERRITORY

To explain how vendors mark and defend their territories, we describe here the process of territorialization and the role of social capital within it. We further explain the extent of territorialization and its facilitation.

The number of people who attend the fair has been increasing; in 2011, more than half a million people participated. As two vendors from Dharmapuri point out: "We come here every year. On these two days, for every investment of Rs.100 we make about Rs.200."

Today, the fair has reached a point of saturation on the Bull Temple Road. According to the manager of the Bull Temple, "The vendors need to pay 100 to 200 rupees per day for occupying public space along the Bull Temple Road. The contract of collecting the rent is given out as a tender, to whoever bids the highest. The temple gets around INR 10,000 to INR 15,000 from it." The bylanes have now begun to be occupied by vendors. Visitors to the fair enter Bull Temple Road from different arteries facilitating new spaces along sidewalks perpendicular to the fair's main spine. There is a spatial hierarchy established on the pedestrianized road with greater importance given to spaces nearest to the entrance of the Bull Temple. While a vendor may not be able to enter the spaces already appropriated near the temple, they can occupy spaces in the bylanes which are the new grounds nurturing new livelihoods. The vendors located here often are not the ones selling peanuts, but products such as toys and plastic items.

Farmers and traders arrive with cartloads of peanuts, as much as a week in advance. They gradually begin to put up their stalls all along the Bull Temple Road. The stalls sell peanuts in the raw, dried, roasted or boiled varieties, as well as temple fair sweets. Other vendors from villages near Bangalore sell traditional toys, clay articles, plastic items, and bangles. Because there is no formal process of allocation of space (Figure 12.2), the vendors' first task is to mark their territory.

A peanut vendor from Banashankari in Bangalore says: "We come a little early before the *Parishe* and mark out the space with lines in paint and every year we sit here at the same spot." Another vendor explains: 'The first time when I came here this space was empty so I put up my stall here. And, every year since then, I've taken the same space." A vendor from Tamil Nadu says "We come early enough to save the place. We put up gunny bags, ropes etc to make space for our goods here." In the interviews, the question "How does space allocation work here?" brought an answer that repeated itself from one vendor to another: "We came here first and therefore this space is ours."

The territory of each vendor (Figure 13.2) moves inwards towards the compound walls of houses and buildings that define the edge of the urban space. The compound wall becomes a rack for hanging items like clothes and plastic bags visually changes the backdrop of the street for these two days. The path of a pedestrian runs between a tree trunk and a two-wheeler; between a lamppost and the heaps of peanuts; between a bamboo pole holding a makeshift roof and a wooden display platform. The path meanders. The urban space is produced differently, a space that is an active phenomenon.

After the territory has been acquired there are different ways in which the vendors defend it. The temple authorities do not keep a written record of who can occupy what place every year. However, the priests have informally known some of the vendors and their families for many years. This informal familiarity provides an extra level of status to these vendors, which in turn helps them get access to prime locations closer to the temple. A vendor from Salem in Tamil Nadu says: "For 10 years we've been sitting here only. We don't allow anybody else to come here. We come here first, to 'catch' the

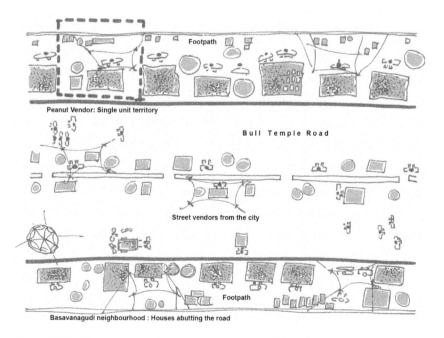

Figure 13.2 The territory of peanut vendors in the fair today.

place." A peanut vendor from Kolar comments: "This place has been with us since my grandfather used to come here. We've always sat here. This is our place and it is noted with the temple authorities that this is our space. They won't let anyone else take it, if they do we tell them it's ours and they have to move." The people decide how the territorialization works best, an informal set of rules govern the 'catching' of place. The catching of place suggests that as people and situations change from one year to the next nothing is ever truly settled, despite the informal rules.

SITUATIONAL SOCIAL CAPITAL

There is little that a vendor wants from other vendors other than to help him occupy a space they may or may not have been using for years, defuse altercations (if there is a contest for the space), and help them possibly transfer the informal 'rights' to siblings or offspring. To enable these activities, collective memory plays a role. It allows vendors to claim spaces and stabilize claims to space at the fair. It becomes possible for a vendor to appropriate a given space because his fellow vendors remember that they have previously occupied the same space or that the residents acknowledge they have bought peanuts from them in the past.

Collective memory arises out of communities that have large amounts of social capital (Bodin, Beatrice, & Henrik, 2006). Social capital is the benefit that groups or individuals receive by virtue of their ties with others (Portes, 1998). For communities to function effectively, trust plays an important role in the formation of social capital (Putnam, 1995). Effective forms of trust and concomitant social norms are likely to arise in communities that live close together and which have a history of interactions with each other (Coleman, 1988).

In the case of the fair the vendors meet for a few days. Many of them do not meet otherwise and meet again only at subsequent fairs. Considering that some of these vendors have been attending the fair for generations, one would assume that the history of interactions can lead to strong social capital. However, in reality when the vendors meet they are quite busy operating their individual ventures. The time and energy spent in interacting with each other is limited, so the possibilities of developing strong social capital are minimal. Nonetheless, during the fair the vendor community displays trust and norms. How is it that an ephemeral community develops such collective dynamism?

In their chapter on temporary groups, Meyerson, Weick, and Kramer (1996) explain that issues small, temporary groups have to overcome may be tied to vulnerability, uncertainty, and risk. The authors argue that if these issues are not managed the participants act more like a permanent crowd than as a temporary system. They define the trust that develops in these groups as "swift trust." Here, the members presuppose trust. They know they are trustworthy and assume they can trust others.

We call this collective collaboration that results from swift trust in a temporary group 'situational social capital.' Situational social capital demonstrates the same characteristics as social capital more generally, except that these dynamics vanish once the group disbands and only return when the group forms again. While it is true that in any group social capital will decay if the members do not interact, such social capital decay tapers slowly (Burt, 2002). However, with situational social capital the trust and norms only exist during the situation.

The situational capital is tied to temporality. In understanding territoriality in the context of the fair we find that space also emerges as a temporal construct. Mathur and D'Cunha (2006) suggest that in the past, as people crossed a land, they imposed on it a new imagination, but also experienced a transformative power exercised by the land. Consequently, they see the "landscape as a phenomenon that is always in the making—both, in the eye and in the land" (p. 214). Doreen Massey, in her book *For Space* (2005), refers to how space is interpreted as being "crossed over" and "conquered." This makes space seem like a surface, continuous, and given. Another approach to space is to interpret it as that which you go through or inhabit. The approach Massey advocates is one where space emerges from social, psychological, and material relationships. That is, space *happens* and has *agency*.

As we have shown, the space of the fair is never settled and despite there being degrees of formality, the informal dimensions—the marking out of space and the periodic occurrence—result in deterritorializations and rearticulations. The informalities of the fair mean space happens differently and different agencies emerge. Municipal regulations and traffic regulations are subjugated for a short period of time; for example, that roads are only for automobiles. In the periodic 'catching' of street space the peanut vendors contribute to the production of a marketplace environment that helps them sell more goods, while they simultaneously change its visual landscape. Further, the collective memory is invoked here once again when the vendors, residents, and customers all say: "The fair came here first and therefore it belongs here." The fair provides a sense of 'temporal anchoring' in a city of changing identity, which is turning from being provincial to cosmopolitan and from being India's 'garden city' to being its 'IT capital.'

Collective memory is again invoked. Research studies on collective memory and urban space note that memories are made visible on the landscape directly from people's decisions and actions embedded within socio-spatial conditions (Rose-Redwood et al., 2008). For the local residents, the fair becomes an urban memory that contributes to the evolving neighbourhood and its relation to the city. This marking of territory at the city level takes place and is defended on the same rationale of collective memory that assigns space to the vendor.

CONCLUSION

An informal capturing of space at the fair shifts and challenges different users' formal rights over a space. The formal shop owner's right to space in front of his shop shifts. The pedestrian's right to space on the footpath or sidewalk shifts. The resident's right to space in front of the house entrance shifts. These spaces are rearticulated. Thus, we find that territoriality in this urban space is as much about the collapse of boundaries between the informal and formal as it is about the emergence of these boundaries in new ways. The periodic change leads to changes in economic and social flows. Formal planning of the city continues to imagine space as a surface on which we are placed and makes little or no connection of space with time or with the activities that take place as an outcome of how space *happens*. In the future, scholars could explore the underlying mechanisms that enable the everyday and periodic practices of space and what they do rather than simply trying to define what they are and so limit their possibilities.

ACKNOWLEDGEMENTS

An earlier version of this chapter was presented at the International Seminar on 'Urban Visualities' held at Dakshinachitra, Chennai, in January 2011.

REFERENCES

Altman, I. (1975). The environment and social behavior: Privacy, personal space, territory, crowding. Monterey, CA: Brooks/Cole.

Arefi, M., & Meyers, W. R. (2003). What is public about public space: The case of Visakhapatnam, India. *Cities, 20*(5), 331–339.

Bodin, O., Beatrice, C., & Henrik, E. (2006). Social networks in natural resource management: what is there to learn from a structural perspective? *Ecology & Society, 11*(2).

Bromley, R. J., Symanski, R., & Good, C. M. (1975). The rationale of periodic markets. *Annals of the Association of American Geographers, 65*(4), 530–537.

Burt, R. S. (2002). Bridge decay. *Social networks, 24*(4), 333–363.

Coleman, J. S. (1988). Social capital in the creation of human capital. *American Journal of Sociology, 94*, S95–S120.

Hall, E. T. (1969). *Hidden dimension.* New York: Anchor Books.

Massey, D. (2005). *For space.* London: Sage.

Mathur, A., & D'Cunha, D. (2006). *Deccan traverses: The making of Bangalore's terrain.* New Delhi: Rupa.

Meyerson, D., Weick, K. E., & Kramer, R. M. (1996). Swift trust and temporary groups. In R. M. Kramer & T. R. Tyler (Eds.), *Trust in organizations: Frontiers in theory and research* (pp. 166–195). Thousand Oaks, CA: Sage.

Murray, J. (1924). *Murray's—A Handbook for Travellers in India, Burma & Ceylon* (11th ed.). London: Murray.

Nair, J. (2005). *The promise of the metropolis: Bangalore's twentieth century.* New Delhi: Oxford University Press.

Portes, A. (1998). Social capital: Its origins and applications in modern sociology. *Annual Review of Sociology, 24*, 1–24.

Putnam, R. D. (1995). Bowling alone: America's declining social capital. *Journal of Democracy, 6*(1), 65–78.

Rose-Redwood, R., Alderman, D., & Azaryahu, M. (2008). Collective memory and the politics of urban space: An introduction. *GeoJournal, 73*(3), 161–164.

Sacks, R. D. (1986). *Human territoriality: Its theory and history.* Cambridge: Cambridge University Press.

Stine, J. H. (1962). Temporal aspects of tertiary production elements in Korea. In F. R. Pitts (Ed.), *Urban systems and economic development* (pp. 68–88). Eugene: University of Oregon Press.

Sudhira, H. S., Ramachandra, T. V., & Bala Subrahmanya, M. H. (2007). City profile Bangalore. *Cities, 24*(5), 379–390.

Weber, A. (1929). *Theory of the location of industries.* Chicago: University of Chicago Press.

Part III

Service, Governance, and Policy

14 The Politics of Space in the Marketplace

Re-placing Periodic Markets in Istanbul

Asli Duru

INTRODUCTION

In Istanbul, food is essential for social connectivity and interaction. Everyday food settings in the city are rich and diverse, ranging from open-air periodic markets and street vendors to transnational super/hypermarkets, and from imported food in high-scale restaurants to boutique grocery shops and foodstuffs transferred individually or collectively from rural areas (Kaldjian, 2003). For most households, periodic markets are an essential resource that primarily provide access to fresh produce, but also to clothing, shoes, and other goods. They join individuals to places and temporal cycles on regional, national, and international scales. Periodic markets are, however, more than a component linking food to the built form and place (Heynen, Kaika, & Swyngedouw, 2006; Swyngedouw & Heynen, 2003). They are subject to the neoliberal political economy of Istanbul, and therefore provide a venue for analysing the context-specific unfolding of "actually existing neoliberalism" in individual lives (Brenner & Theodore, 2002, p. 349).

Current changes in the organization of Istanbul's periodic markets involve, above all, displacement and "downsizing" (Oz & Eder, 2012, p. 298) to remote and less accessible locations by municipal governments and through national government incentives. Peripheral displacement of periodic markets reflects asymmetries in access to, and control over, the key material resources for urban livelihoods: food and space. Given that there has been no significant decrease in demand for markets in core urban areas where spatial commodification is most intense, municipal relocation of periodic markets suggests profit-driven reorganization of urban resources. Transformations in the mobility and practices of urban subjects (residents, vendors, shoppers, visitors) that are derived from current entrepreneurial modes in municipal governance of markets do not take into account the lives and livelihoods of present urban subjects. Instead, they reorganize inclusion and mobility within relationships between state and individual bodies for the not-yet-accomplished potentialities of future subjects and space.

In this chapter, I present a layered, subject-centred approach to the changing sites, patterns, and perceptions of markets in Istanbul, and thereby address the multiple challenges experienced by residents who are core providers of food-related care in their household. A perspective informed by gender and livelihood guides the research to explore the impact of periodic market relocations on the perceptions and practices of everyday and long-term mobility for women in their roles as caregivers and provisioners for their families. This is assessed within the context of an urban resource base subject to neoliberal regulation. My reference to neoliberalism positions the urban transnational flows of people and goods as a historical, political, and economic process that generally entails restrictions in formal and informal social and material assemblages of services and resources, such as periodic markets.

The study is based on 40 food biography interviews carried out in 2010 and 2011 with provisioning women who are affected directly or indirectly by the transformations in the periodic market network in their area. As a research tool, food biography recognizes the centrality of food (Cook, Crang, & Thorpe, 1998, p. 166) in constructing intimate and social environments. While there were no strict selection criteria with regard to specific aspects of subject profiles such as age or income, participants were sought on the basis of their role as the primary provider of food-related care within their household.

PERIODIC MARKETS ISTANBUL

The market has long been a rich combination of routes, formal and informal institutions, and modes of exchange that meet the long- and short-term social and material needs of Istanbul's population. Following the Ottomanization of the city in 1453, Istanbul grew in population and area, and periodic markets were established in the areas surrounding the city walls, some of which, such as the Wednesday Market in Fatih, continue until the present day (Goktas, 1997). Most central periodic markets reached their present form (in size and location) and significance in the 1970s, following massive waves of rural migration. Despite the changes in their location and organization over time, periodic markets have remained important sites for access to food and other everyday resources, and for social interaction. Among the variety of markets in Istanbul, weekly open-air markets are the most widespread network contributing to the urban food supply (Oz & Eder, 2012).

Typically, a market is held on a certain day of the week in a particular neighbourhood and is then most commonly referred to by the name of the day on which it is held, such as the Tuesday Market (Salı Pazarı in Kadıköy) or the famous Wednesday Market (Çarşamba Pazarı in Fatih). Markets are usually set up in a network of designated streets and/or lots, where stalls are grouped according to the type of goods sold. Periodic markets are co-organized and administered by district municipalities and city-based vendor associations (Chamber of Periodic Market Vendors, or CPMVs), and the public lands on which they are held may belong to a district, a metropolitan municipality, or to the National Treasury. According to the Istanbul CPMV data in 2013, there are 377 periodic markets operating in Istanbul on a weekly basis. Neighbourhood periodic markets are a flux of formal arrangement and informality. Although the goods sold are more or less the same in many periodic markets, the price range and quality of products are dependent on population density and the socio-economic level of the neighbourhood (Dokmeci, Yazgi, & Ozus, 2006). Geographical variations enable the price flexibility and product diversity that are vital for livelihood provision to diverse socio-economic groups. The semiformal setting in which social interactions between sellers and buyers take place is more participatory (Williams, 2002), and therefore more appealing and empowering for customers. The formal organization of the spaces and of the terms of practice, together with unregistered vendors and bargaining, make the market a uniquely flexible system of commercial activity. Informality benefits sellers because the "small and labour intensive trade units" (Dokmeci et al., 2006, p. 46) are mobile, can set up in a different market each day of the week, and can make up for losses in a less profitable market by also selling in a better-off neighbourhood.

It is difficult to quantitatively represent periodic markets' present share within the urban food supply, or how that share has evolved over the long course of market activity in Istanbul (Uzuncarsili & Ersun, 2004). Indeed, research on periodic markets in Turkey is scant compared with the literature on markets in rural and urban Latin America, South Asia and Africa, or farmers' markets in North America (Holloway & Kneafsey, 2000; McCormack et al., 2010; Lamb, 2010; Brown, 2002). In Istanbul, periodic markets are mostly referred to as part of the wider retail network undergoing rapid transformation as a result of global trends, the inclusion of new retail forms, and the spread of new taste and consumer profiles. Dokmeci and colleagues (2006) address these trends in relation to the distribution of commercial space in Istanbul in the period 1980 to 2002. Their findings indicate that while the number of markets emerging in the city's newly developing neighbourhoods grew due to increased population density, established markets have been displaced from central areas of Istanbul despite there being no significant change in population density.

In the last three decades urban space in Istanbul has been marked by the discourses and priorities of intensive capitalist market forces. Rent-oriented place-making and gentrification of core areas inscribe the cityscape with the priorities of new growth paradigms. Concomitantly, the social base of urban politics is mobilized in terms of the future interests of private investment and middle-class consumerism through strategies such as political populism or zoning and census management, which target, maximize, or reorganize groups in working-class neighbourhoods. An example of the former is the discursive justification of the displacement of existing formal and informal periodic market spaces on the grounds of security concerns

(Dikec, 2007). A comprehensive examination of urban manifestations of state-subject relations must address the relocation of periodic market locations given that these changes entail the reorganization of everyday practices of provision and of residents' individual mobility patterns. Periodic markets are therefore an entry point for critically examining both micro and macro processes in the making and remaking of the urban in Istanbul.

PERIODIC MARKET REGULATION AS FORMAL SPATIAL STRATEGY

The modernization of infrastructure is an officially stated objective of the relocation and redevelopment of periodic markets. Markets with acceptable infrastructure exist only in a few locations in well-off neighbourhoods. Investment in the infrastructure of the market setting is favoured by sellers, primarily for reasons of customer satisfaction. Camera surveillance, availability of air conditioning, non-cash payment options, and parking spaces signify a modern and streamlined market experience for consumers, yet projects to improve substandard market environments mostly override vendor and customer interests. In addition to forcing onerous travel conditions on shoppers, the move to peripheral, remote locations denies vendors the flexibility to move between centrally located markets and imposes a heavy financial burden on the operating costs of small-scale traders.

The regulation and locating of periodic markets have until recently come under municipal jurisdiction. However, the role of the national government in the regulation of periodic markets was redefined in the By-law on Market Places issued by the Ministry of Customs and Trade in 2012. The by-law aimed to stop divergent applications of individual municipalities when reorganising the periodic markets within their administrative boundaries. It was claimed by the ministry that the previous situation led to dissatisfaction on the part of both market traders and consumers. Before it was put into effect in July 2012 the draft by-law was made publicly available in 2011, and the ministry reported ongoing, nationwide consultation with stakeholders, including trade organizations and wholesalers. From the vendors' perspectives, the drafting of the new regulation was hardly transparent as the ministry was not responsive equally to all stakeholders (per the author's informal conversations with vendor/activists). Public discussion of the changes was largely limited to the interests of large-scale agricultural traders, and the newly introduced legal framework was motivated by a desire to centralize the downsizing of municipal periodic markets (Pazarci, 2003; Oz & Eder, 2012).

Periodic market governance can be understood as part of the wider transformation of politics in Istanbul, and as such represents the spatial manifestation of "actually existing neoliberalism" (Brenner & Theodore 2002, p. 349). A key feature of neoliberal governance is the integration of economic efficiency as a parameter for politicising municipal governments and their actions. Periodic market relocations are discursively linked to this premise in two ways, both of which reflect dimensions of a pro-growth agenda in the regulation and management of space as an urban resource. First, periodic markets in central lots/streets—in that they are economically informal and aesthetically 'unappealing'—reflect an 'inefficient' use of valuable urban space. They are relocated in order that public space can be economically potentialized through the formal economy, often through private investment. The second, complementary motive for relocations is the removal of informality and visible poverty from central Istanbul (Oz & Eder, 2012). Both objectives promote spatial and social disparity and exclusion, which translate into further limitations on the everyday and longer-term provision of livelihoods and food for residents.

PERIODIC MARKET SHOPPING AS GENDERED SPATIAL STRATEGY

In the context of Istanbul's intense political, economic, and physical transformation, periodic markets comprise the ground for examining what these changes signify in terms of the spatial and temporal arrangements of everyday mobility for residents. The location of a periodic market is a defining element in negotiating everyday individual mobility because it has a central role in the spatial organization of accessing domestic provisions. Residents from various socio-economic backgrounds are attracted to periodic markets for different reasons, but for less mobile and less well-off provisioners, accessibility and availability of inexpensive, varied produce are key.

Food biography analysis is attentive to recurring performative, mental, and emotional patterns in everyday provisioning routes and practices. As people talk through their experiences, they construct the representational realm that unfolds through their accounts. Speech promotes a subject's capacity to dynamically design, produce, and reproduce lived experience. Food biography as a method recognizes this aspect and integrates the subject's own engagement with place, time, distance, and chronology. Thus, it is also concerned with the narrative sequence and organization of information transmitted during the interviews. The research process in this study entailed sequential biographical interviews with female care providers who self-identified as 'mid-range,' an economic category which broadly indicates a self-sufficient household socio-economy. 'Mid-range' economic status is an extremely vague category, and within it, participant profiles diverged significantly in terms of the material and social conditions that characterized their livelihoods. Thus, although the provisioning women in the study belonged to the same economic category, they were diverse in terms of age, education, household and employment status, ability, ethnicity, income, and availability of other material and nonmaterial assets. There is also significant variance in terms of provisioning women's family structures, because the people for whom they provide care are likewise diverse. Women's lifestyles are shaped by factors such as their socio-economic circumstances, their geographic location, and their multiple responsibilities as mothers and employees.

In all of the food biographies, whether regular market shoppers or not, women acknowledged the availability or proximity of a market as an important element in ensuring sustained access to healthy and fresh food resources. Such availability ensures that permanent food retailers such as supermarkets and neighbourhood shops remain price competitive.

In the study, all subjects responsible for provisioning considered periodic markets to be a significant part of their personal foodscapes and everyday mobility, and were aware of the changes taking place at markets in their neighbourhood. More than half the food biographies of those I interviewed (25 out of 40) highlight the significance of having a weekly market in their neighbourhood, not only as a public service for limited-resource households in a locality, but also as a factor that creates a collective benefit for the wider neighbourhood population. All subjects regarded the relocation of a market to a permanent site in a central location as a positive change. On the other hand, the displacement of a market to the peripheral streets or areas of a neighbourhood was the main reason 13 of the participants gave for no longer shopping there. Displacement is usually compensated by shopping at small- to medium-scale supermarkets that proliferate in central areas (such as central roads and streets where weekly markets were previously set up) and offer 'public day' deals, although with a lower diversity and quality of produce.

During our meetings, the descriptive sketches of different periodic markets and the women's use and perception of directions, routes, distance, and past and present mobilities were most often visualized as nonlinear, complex cartographies with highly subjective mobility patterns of primary, secondary, and tertiary (and sometimes more) route options. Seventeen of the 40 food biographies declare walkable distance as the most influential factor determining their assessment of a periodic market's accessibility. For all of the subjects who were in paid employment, walkable distance within the route of their daily commute was expressed as a strategy, in order to benefit from the lower prices of an evening periodic market on the way home from work. Significantly, the mobility required to access a certain food site is factored into the affordability of produce from a certain source. Distance to an available periodic market, according to the women's accounts, is calculated by the "optimal distance that can be travelled by walking" (respondent 27) and in most cases without "having to go too far from home" (respondent 14). Women described a certain periodic market as available when it is within a distance that can be travelled "cost-free" (respondent 14) and "without spending too much physical energy and time" (respondent 35). In all cases, women's mobility planning involved intense strategising over time and space.

The women managed layers of sense-based and informal knowledge dynamically on a weekly basis. Based on the accounts, the variables I identified in this assemblage of motion, affect, and interaction relate mostly to the informal and subjective contingencies such as distance versus time saving, atmosphere of the market, family budget and preferences, physical ability, and factors such as weather, vendor interaction, and therapeutic features of distinct periodic market visits. Longer walkable distance is by far the most influential factor for no longer shopping at a former centrally located periodic market. Women shopping at a market describe their mobility on a market day as designed to achieve an efficient use of time, money, and energy, thus relocation of markets from central locations to peripheral areas in a neighbourhood prompts a sharp redefinition of the everyday mobility of women. Extended travel time to a relocated market is considered "not worth it" (respondent 17), "physically tiring" (respondent 14), "sickening and inconvenient" (respondent 21) and in some cases "impossible" (respondent 2). Distance to a periodic market conceived in terms of time and physical burden of the activity is a factor in women's calculation of the quality–affordability continuum. For 19 out of the 29 shoppers who lacked both time and stable income, weekly market shopping reduced their reliance on other sources (primarily, supermarkets) to optimize quality and affordable food throughout the week. This was a recurring feature in the mobility patterns of women caring for larger households, including young children and older parents, when they commented on their transformed access to a relocated market. These subjects relied more on centrally located periodic markets to plan affordable, adequate, and preferred meals, with relatively limited resources, including, above all, time, income, and physical ability.

This equilibrium is usually an experiential product of women's expertise and informal knowledge in balancing individual and household resources with perceived needs, to achieve the "healthy, happy, and productive" (respondent 12) functioning of their household and "peace in the family" (respondent 12). Therefore, changes to the spatial, social, and temporal organization of periodic markets relate with varying intensity and for different reasons to the performance and conception of identity for provisioning subjects. In the case of older women, the reorganization of periodic markets represents a significant limitation on everyday mobility patterns and provisioning capacity. For younger subjects with limited resources, the new locations mean loss of formerly available food sources and/or longer travel distance, as well as a limited personal foodscape to conceptualize individual provisioning strategies. The relocation and displacement of neighbourhood markets influences women's everyday mobility in material and nonmaterial ways. These changes affect the family provisioning equilibrium and shift it in directions that are sometimes beyond the capacities of aged, mobility challenged, and/or economically deprived family provisioners.

156   *Asli Duru*

CONCLUSION

This project grew out of personal observations on the significance of food provisioning in periodic markets in Istanbul. The theme developed as a means of thinking about urban governance and the constitution of subjects through the material and social regulation of space. In Istanbul, municipal and national reappropriation of urban land and the commercialization of public space represent a banal, globally ubiquitous policy of promoting growth through commodification of urban resources. In the case of the city's periodic markets, such policy overlooks the implications displacement has for the livelihoods of some subjects. To make sense of how urban spatial strategies are materialized in subjects' practices and perceptions of place, my focus has been on provisioners' social and spatial relationship with Istanbul's periodic markets. Such an approach to place and politics is relational, prioritizes subjects, and relies on the thick information and the informal knowledge gleaned from people's narratives and language.

A closer look at the present assemblage of formal and informal forms of spatial organization, and social, material, and knowledge exchange in markets suggests the reasons why they present a strategic alternative to financialized, intensely regulated provisioning places such as supermarkets and certified organic farmers' markets. They are important for residents whose livelihood strategies and access to food rely on the marginal affordability of fresh produce in the market. Counter to municipal and national government discourse regarding the value of markets, they are also an efficient use of urban resources compared with retail in the formal economy. Relocation and/or closing of weekly markets maintains a hierarchy of urban priorities that overlooks and erodes the potential in the informal use of urban space and in informal, sense-based, unregistered forms of urban knowledge. In the case of Istanbul, narratives of and from provisioning women present the ground for tracing informal strategies and responses that attempt to reclaim individually efficient mobility patterns and control over livelihoods. On a broader analytical level, this kind of informal knowledge affords critical insight into the state's repositioning of subjects within the context of Turkey's contemporary developmental momentum, and the crucial implications for women's 'right to the city' in terms of mobility and livelihood.

REFERENCES

Brenner, N., & Theodore, N. (2002). Cities and the geographies of "actually existing neoliberalism." *Antipode, 34*(3), 349–379.
Brown, A. (2002). Farmers' market research 1940–2000: An inventory and review. *American Journal of Alternative Agriculture, 172*(4), 167–176.
Cook, I., Crang, P., & Thorpe, M. (1998). Biographies and geographies: Consumer understandings of the origins of foods. *British Food Journal, 100*(3), 162–167.
Dikec, M. (2007). *Badlands of the republic: Space, politics, and urban policy.* Oxford: Blackwell.

Dokmeci, V., Yazgi, B., & Ozus, E. (2006). Informal retailing in a global age: The growth of periodic markets in Istanbul, 1980–2002. *Cities, 23*(1), 44–55.

Göktaş, U. (1997). Eski İstanbul'da Semt Pazarları. *İlgi, 90*, pp. 8–11.

Heynen, N. C., Kaika, M., & Swyngedouw, E. (Ed.). (2006). *In the nature of cities: Urban political ecology and the politics of urban metabolism.* London: Routledge.

Holloway, L., & Kneafsey, M. (2000). Reading the space of the farmers' market: A preliminary investigation from the UK. *Sociologia Ruralis, 40*(3), 285–299.

Kaldjian, P. (2003). Urban food security, the rural hinterland, and Istanbul's lower income households. In Hans Löfgren (Ed.), *Food, agriculture, and economic policy in the Middle East and North Africa,* Research in Middle East Economics, Vol. 5. Oxford: Elsevier.

Lamb, D. (2010). *Scottish farmers' market research.* SAC Food and Drink.

McCormack, L., Laska, M., Larson, N., & Story, M. (2010). Review of the nutritional implications of farmers' markets and community gardens: A call for evaluation and research efforts. *Journal of American Dietetic Association, 110*(3), 399–408.

Oz, O., & Eder, M. (2012). Rendering Istanbul's periodic bazaars invisible: Reflections on urban transformation and contested space. *International Journal of Urban and Regional Research, 36*(2), 297–314.

Pazarci. (2003). *AB Uyum Sureci.* Retrieved from www.istanbulpazarcilarodasi.com/dergi-ab-uyum-sureci-38

Swyngedouw, E., & Heynen, N. C. (2003). Urban political ecology, justice and the politics of scale. *Antipode: A Journal of Radical Geography, 35*(5): 898–918.

Uzuncarsili, U., & Ersun, O. (2004). *İstanbul'daki Semt Pazarları Envanter Çalışması.* Istanbul: ITO Yayınları.

Williams, C. C. (2002). Why do people use alternative retail places? Some case study evidence from English urban areas. *Urban Studies, 39*, 1897–1910.

15 Shanghai's Unlicensed Taxis (*Hei Che*) as Informal Urban Street Market

Dunfu Zhang

Hei che (黑车) refers to an unlicensed taxi, also known as a 'black taxi.' *Hei che* are common throughout China, from small towns to global metropolises such as Shanghai, Beijing, and Guangzhou. Through an ethnography of the *hei che* phenomenon in Shanghai, I argue that rather than the stigmatized 'trouble-maker' or illegal operation, *hei che* can be understood as an informal urban street market where self-employed rural-urban migrant workers strive to make a living in the context of the Hukou System. This system of household registration determines whether someone can work in a place, or have access to housing and social services, such as education and hospitals, in the city where they are currently working. Many rural to urban migrants (*nongming gong*) do not have a Shanghai Hukou card, as they are very difficult to obtain. Shanghai is a leading global city, and was the leading "dragon head" (*Longtou*) during China's reform era. As Lucian Pye (1981) explains, "serious analysis of nearly all important aspects of life in China must, eventually, confront Shanghai and its special place in the Chinese scheme of the things" (p. xi).

There is a large 'floating people' of China, predominantly rural-urban migrants. Many enter into the informal economy, particularly by participating in informal urban street markets. This population's production, labour and consumption are a growing part of China's informal economy in urban areas and signal this economy's social and economic significance, as well as its integration with the formal economy (Huang, 2009; Light, 2004; Castells & Portes, 1989). This chapter interrogates Shanghai's informal economy and informal urban street markets through an ethnography of *hei che*. To date, no research has been done at the street level in regards to this informal urban street market. Without such research there is misunderstanding, misinterpretation, and mistreatment of *hei che* drivers and their customers in Shanghai, and subsequently a lack of evidence-based policy that could facilitate better relationships and equitable practices.

This chapter is framed by three objectives. The first is to provide an in-depth description of the *hei che* in Shanghai, including the discourse that frames *hei che* as being a 'headache' for governmental regulation. The second is to consider how *hei che* provide a means of livelihood for less privileged people, such as *nongming gong*. Further, I will explain the economic and social significance of *hei che* for consumers. Finally, this chapter will look at how these factors relate and play out as a particular informal urban street market that is put under strain by a powerful formal authority that aims to curtail operations. To be more specific, district-level governments throughout Shanghai work hard to prevent *hei che* from operating.

Most of the empirical data used in this study was collected through participant observations while riding in *hei che*. I conducted interviews with differing degrees of formality, including unexpected encounters, prepared questions, and group interviews. As part of the encounters I employed 'guerrilla interviewing.' According to sociologist Thomas Gold (1989), this is "unchaperoned, spontaneous but structured participant observation and interviews as opportunities present themselves" (p. 180). I used this strategy because participants were at times reluctant to talk in front of others, and because what they had to say about crime, bribery, and corruption could get them in trouble with the law. It was very hard to establish trust and persuade drivers to cooperate with researchers in a formal way. They prefer to keep their business private from outsiders. Inquisitive passengers or strangers are sometimes suspected of being spies of the police or government. However, guerrilla interviewing worked well as participation took place in their space (e.g. inside the car) and when they wanted it. The journeys were of variable length and frequency depending on how familiar I became with the drivers. Given these conditions, the interviewees relaxed and trust was established.

A reason trust is hard to establish is that *hei che* customers and drivers are squeezed by powerful institutions and cultural discourses. Before one gets to know unlicensed taxi operators, what you hear about *hei che* are stereotypes, characterized by adjectives such as dangerous, cheating, violent, and deviant. One online tip explains,

> Shanghai has the best-managed taxi service in China . . . Licensed taxi means it is necessary to have a meter and an illuminated vacancy disk on the dashboard. Without all these things, the taxi is probably unlicensed and you should avoid it, even if the driver solicits you. You have no rights if injured in an unlicensed taxi.[1]

This advice from a popular website is provided to both foreigners and Chinese nationals. Reports of cheating, unjust fare charges, and robbery are frequently published in newspapers and via the popular web portals sina.com and sohu.com. It has been claimed that one *hei che* driver drove 70 kilometres to the passenger's destination, which was only eight kilometres away, and charged 10 times the reasonable price.[2]

News website Shanghai Online reported that a *hei che* driver killed a female passenger and disposed of the corpse after an unsuccessful rape.[3] On July 4, 2011 the *Dongfang Daily* newspaper had an article about how Songjiang's most popular traffic policeman Qian Yong was attacked by a *hei che* driver.[4] An expatriate health website instructs its customers that "To avoid such frequent instances of sexual harassment, expat women may prefer taking a taxi. With the sole exception of the odd unlicensed cab."[5]

Slogans such as "For your safety, please do not take *hei che*" are often seen at popular transportation centres. Minhang District Government in Shanghai issued two separate public letters warning about *hei che*. One letter pressed drivers to give up the illegal business, and the other asked potential passengers to avoid and report the use of *hei che*. In 2012 my son received a formal public letter via his school warning about *hei che* operation and use. The letter was co-issued by the Minhang Educational Committee and Shanghai Minhang District Traffic Administration Enforcement Division, and was sent to all students' parents. The letter stated that the "[i]llegal operation of non-Shanghai license plate *hei che* adversely influences the normal business of legal taxis and harms social order." One of my informants, a newcomer to Shanghai, told me:

> About six or seven years ago I went to Pudong airport to see my friend off. The next day I planned to go back home. My friend gave me a card telling me to call the man whose telephone number was on it. The driver was to take me to HongQiao station, where I was supposed to board a train. The man asked me questions at times and I replied reluctantly until something unpredictable happened: the man dropped me off half way to the destination. Actually, I was forced to get out of his car without any explanations or apologies. I paid him 60 or 70 Yuan. He could definitely tell I was a stranger in Shanghai because of my dialect and conversation.

On March 20, 2012, I saw a car with an Anhui province badge stop at a crossroads in my town of Zhuanqiao, Shanghai. Anhui is a province next to Shanghai. The driver tried to drag a young girl out of the car while a young man stood by with three pieces of luggage. It seemed they had agreed to take the ride yet disagreed with the driver about the price. When I tried to approach to see what was happening the driver threatened me: "What for? Go away!" My cousin's experience is similar.

> I came to Shanghai with my wife, hopefully to see an expert doctor to cure my liver disease. You called me to take the formal (*zhenggui*) taxis to get to your apartment. When we got out of HongQiao station, a young lady approached us: "Taxi, taxi!" I thought they were the formal ones, so I followed her to a car. The middle-age driver gave us the ride for 100 yuan. Now you told me the legal taxi driver will charge about 70 yuan for the same ride.

Hei che fares are often more expensive than formal taxi fares. I asked one familiar driver if he could take my friend to Shanghai's Pudong International

Airport from my apartment for the same price as a formal taxi, which would have been 200 yuan. His answer was, "Sorry, big brother, if only you could pay more, around 300 yuan. You know we cannot pick up passengers at the drop-off point like (legal) taxis."

However, there are also drivers who will charge reasonable prices. This happens for familiar and regular passengers. One user of *hei che* told me

> I could not see my guest off to the airport. A taxi is not easy to get here. So I gave her one name card of a *hei che* driver, "Anhui Fatty" is what we call him. He's a friendly and honest guy. From our dormitory to Wujing town to the airport a taxi charges 10 yuan for a single trip. Anhui Fatty said 15 yuan is fine for the round trip. I have three *hei che* name cards.

Sometimes a friendly atmosphere exists when riding in a *hei che*. Professor Zhou (a visitor from the city of Guangdong) took a *hei che* with his two friends, chatting with the driver during the journey. When Zhou asked the driver how much, the reply was, "you name it." Zhou paid 20 yuan, and both parties were happy with this. Zhou's local friends told him that the price was good for such a distance on a rainy evening. Price depends on who gives the ride and the conditions of that journey. The two parties often discuss the price to reach an agreement. One passenger suggests that the negotiable price is a *hei che* advantage because "*Hei che* can sometimes be cheaper than licensed taxis, and more convenient. We can bargain with the driver for a good price."

Some *hei che* operators have established strong trust between driver and regular passengers because of the lack of transport alternatives. One customer told me,

> Our company is located downtown. We take *hei che* rides quite often, but we often use the same driver. The driver used to work for a taxi company. We are now familiar with each other. I won't take other *hei che* no matter who they are and how cheap the price is, not even a formal taxi. This driver gives us the same receipts as the big taxi corporations. We feel secure and safe riding with him.

Hei che are popular for those who need a ride when formal taxis are not available. Often the last few kilometres of a journey in some areas of Shanghai are unmet by public transportation. People use *hei che* because they are necessary. This is especially the case in newly developed suburban areas when public transportation finishes at around 10:00 p.m. or when the Metro and public bus systems do not coordinate well. As one passenger told me, "I don't want to take *hei che*, but I have no other choice."

Hei che are often found in outer and newer suburbs of Shanghai, such as Nanhui, Fengxian, Minhang, Songjiang, Jinshan, Jiading, Baoshan, Qingpu, and Pudong New. These areas are identified as the 'less-civilized,' 'dirtier,' 'messier,' and 'shabbier' parts of Shanghai, where migrant workers and relocated middle- and lower-class urbanites find jobs and housing. *Hei che* also gather in manufacturing areas. In July 2010 I undertook fieldwork in Shanghai's Songjiang export processing zone and found dozens of *hei che* near the Foxconn dormitory, a major telecommunications manufacturer that has contracts with Apple Inc. (among others). The young workers explained that where there are few, if any, public transport options *hei che* provide access to downtown commercial centres. Similar phenomena exist in other newly emerging high technology development areas like Pudong or Minhang. The same can be said of new university campuses, especially suburban campuses like the Baoshan campus of Shanghai University, the Minhang campus for both East China Normal University and Shanghai Jiao Tong University, Songjiang University Town, and Fengxian University Town.

However, *hei che* are seldom found in downtown Shanghai. A Fengxian district operator in Shanghai told me, "Such popular places as People's Square and the railway stations are strictly administered, nobody dares to go there, once caught (the driver) will be fined 5000 yuan." Sometimes *hei che* drivers do run the risk of fines, and operate at popular places like hospitals and shopping malls. On March 14, 2013, I saw three men soliciting patients outside Changhai Hospital in central Shanghai. A fruit shop manager nearby told me "they are a fixture here." Formal cabs hold the monopoly on street pickups but are subject to costly regulations and licenses, which are reflected in their fares. The monopolistic limits on the number of taxi licenses, in order to maximize revenues on every shift, have led to a shortage of taxis in Shanghai. Consequently, people respond to unmet needs in the formal economy through informal provision (Light, 2004).

Rural to urban migrants can find it difficult to gain employment in Shanghai, and when they do, the position is often poorly paid. Self-employment as a *hei che* driver becomes a promising possibility owing to the flexibility of the hours and because good money can be earned.

> There's nothing bad about our job. Shanghai is full of so many people. We drive to serve the masses. We give faster and more convenient delivery at a lower price. We often work hard very late night and very early in the morning, the same as everyone here, to make a better living.

Besides business cards, the drivers can get formal receipts from their taxi company connections to attract riders who require such receipts, such as those who work in the public sector, corporations, or education institutions. One driver joined a formal taxi company but he drove *hei che* as his second job. He has a friend who can print the receipts to sell or to distribute among his peers. He told me, "Want receipts to reimburse? Yes, we have these. I have the formal receipts. You can have as much as you want and we will print the details according to your instructions." The drivers have to work hard to make a living. A driver at the entrance of East China Normal University said,

> We just make extra cash to cover gas costs and car insurance. I stay here for half an hour without business. I rent a local peasant's house, for 400 yuan a month, a shabby one for this price. We can't afford apartments. My wife and my child are with me. We both have our own job in the daytime. We drivers have to be very cautious of being caught by the police, once caught we will be fined 200 yuan and three points on your driver's license. You get 12 points per year. Four times caught and your license will be confiscated. I was caught one time last week, bad luck. We will try to slip away, but sometimes it is too late before the police car stops in front of yours.

I was told that drivers play 'hide-and-seek' with the police. One evening at Jianchuan Road station on Metro line 5 at 9:20 p.m. there were no *hei che* because of the presence of a police car. Once the police left the *hei che* arrived.

Most of the drivers choose this work because they enjoy the freedom and benefits that come with driving their own car. One driver has been a *hei che* operator for eight years. He bought a car for 90,000 yuan, together with the license plate, after he quit his job, which paid only 2,000 yuan a month. The monthly gas costs him 3,000 yuan, with other costs equating to 1,000 yuan. His total income is 10,000 yuan, with the profit being 6,000 yuan per month, and with the help of his wife's small shop, he is able to support his two-child family. Like other operators, he makes better money on weekends and holidays. A driver in Jiading district in Shanghai told me,

> There are a few of my friends who joined formal taxi companies but there are so many regulations there that most of us now prefer driving *hei che* since we make more money without paying tax, and use less gas because of our car brands. Yes, the job is unstable but there's lots of freedom. Most of the time you can do whatever you want. We don't like to be supervised by a company boss or an official. We make decisions about our own life.

The informality of this urban street market presents opportunities for agency for a population which has few such opportunities otherwise, especially without a Shanghai Hukou card.

One group of drivers at a popular *hei che* location consists of stable members who know each other well. They chat and joke together when there is no business. Some of them are close friends and have relatives from the same town in neighbouring provinces. As close town-fellows they have developed their own unwritten business rules. More often than not they park in a queue. The second car in the queue can only take passengers after the first one leaves. If a passenger insists on a particular car, and the drivers agree, the regulation may change, especially if the drivers have very close ties and the passenger is a regular one. The informality enables flexibility.

> You can pick any car you like. The car next to mine is my younger brother's. Our hometown is Anhui Shou county. We were brought here one by one, and we bring in new relatives and friends. We know each other from the same town, but not necessarily from the same county. People from the same place know each other quite well, that works much better.

If drivers do not know the other drivers at a pickup location, other informal arrangements come into play. Around 10:30 p.m. one night, when I was curious if a driver could pick up someone from the nearest Metro station, the driver said,

> I have to go back to where I picked you up, since it's impossible to get a call-in customer just here. It's hard to stop at the nearby Metro station near your apartment to wait for passengers. These places are taken (by other *hei che* drivers). I don't know them and it's not right or fair to infringe on their territory.

It is safe to say that often each popular *hei che* location is regulated by a certain group of drivers who have strong ties, and informal rules emerge and are abided by.

Most *hei che* drivers I met in Shanghai come from the nearby province of Anhui. Anhui provides many rural-urban migrant workers to Shanghai, such as construction workers, nannies, property management staff, and beauty salon assistants. The drivers were predominantly in their twenties or thirties, although a few were in their forties, and I was told the most literate had graduated from junior school. Their educational background is one of the reasons they are excluded from professional jobs and have to "play with cars" (as they call it). At Dong Chuan Road station, Metro line No. 5, one driver told me,

> I'm from Fengtai county, Anhui Province. You can tell here the *hei che* plates are all Anhui ones. We all came here after being introduced by *laoxiang* (town folks). We take care of each other. We are not afraid of the police or vehicle administrative staff. We have someone protecting us from above (in government), you know, who let us know ahead of time if a clearing will happen. Even if we were caught our connections (in the government) can help us to get our vehicles back.

To protect them from the police or dangers such as robbery some drivers have established 'pseudo-relatives' by becoming sworn 'brothers,' most of whom are from the same town or county. If there are signals of danger brothers will let each other know. They respond quickly to help any brother in difficulty.

CONCLUSION

The informal economy has grown dramatically in developing countries since the 1970s. However, China's state statistical apparatus continues to neglect the informal sector (Huang, 2009). The informal urban street market of *hei che* that I have described here is an example of how this informal economy emerges and works on the streets of Shanghai. The market is primarily underwritten by rural-to-urban migration, and as such *hei che* are primarily located in areas most influenced by this migration. The rural-urban social distinction perpetuated by the Hukou System is influential because without the correct residency permit card it is difficult to find work and migrants do not get social welfare support. As one driver said,

> Shanghai taxi (companies) are state or municipal enterprises, they are not open to us rural migrant workers, we cannot enter. You have to get Shanghai Hukou here, like those Chongming District guys born to have it. We are not qualified (to have a Shanghai residency/work permit card). We have to use our hands, our own cars to create the jobs and lives ourselves.

In *Social Structure and Anomie*, Robert Merton (1968) offered the insightful proposition that "social structures exert a definitive pressure to engage in non-conforming rather than conforming conduct" (p. 186). Merton focused on deviance and anomie, but he recognized innovation as another kind of nonconformance. Unlicensed taxi operation offers a case that drivers innovate to make a living, sometimes to live a better life. Far from the stereotyped trouble-makers, the unlicensed taxi drivers form an informal urban street market evidencing creativity, such as the organic emergence of cultural rules to share wealth and opportunity. There is agency here. There is thus a need for recognition of this when deciding on city policies. However, the government, together with the formal legal taxi corporations and monopolized mainstream media, continues to discursively frame *hei che* and the drivers as illegal and dangerous, and in doing so oppresses the innovation and the livelihoods of these rural-urban migrants.

ACKNOWLEDGEMENTS

This paper was financed by "Studies on the Social Quality and Construction of Harmonious Society," a key project of the National Social Science Foundation of China (No. 11&ZD148).

NOTES

1. www.zlrealty.com/user_list.asp?bookid=519.
2. http://sh.sina.com.cn/news/s/2012–03–03/1556210025.html.
3. http://news.online.sh.cn/news/gb/content/ 2011–12/07/content_4999791.html.
4. http://epaper.dfdaily.com/dfzb/html/2011–06/04/content_489388.html.
5. www.internations.org/shanghai-expats/guide/living-in-shanghai-15424/transport-and-healthcare-in-shanghai-2.

REFERENCES

Castells, M., & Portes, A. (1989). World underneath the origins, dynamics, and effects of the informal economy. In A. Portes, M. Castells, & L. A. Benton (Ed.), *The informal economy: Studies in advanced and less-advanced countries* (pp. 11–37). Baltimore, MD: Johns Hopkins University Press.

Gold, T. B. (1989). Guerrilla interviewing among the Getihu. In P. Link, R. Madson, & P. G. Pickowicz (Eds.), *Unofficial China: Popular culture and thought in the PRC* (pp. 175–192). Boulder, CO: Westview Press.

Huang, P.C.C. (2009). China's neglected informal economy: Reality and theory. *Modern China, 35*(4), 405–438.

Light, D. W. (2004). From migrant enclaves to mainstream: Reconceptualizing informal economic behavior. *Theory and Society, 33*(6), 705–737.

Merton, R. K. (1968). Social structure and anomie. In R. K. Merton (Ed.), *Social theory and social structure*. New York: Free Press.

Pye, L. W. (1981). Foreword. In C. Howe (Ed.), *Shanghai, revolution and development in an Asian metropolis* (pp. xi–xvi). Cambridge: Cambridge University Press.

16 Geographies of Unauthorized Street Trade and the "Fight Against Counterfeiting" in Milan

Kate Hepworth

Corso Buenos Aires, Milan's high street. From Piazzale Loreto in the northeast and its nameless stores of cheap Chinese-made, unbranded clothes, the road runs southwest past global clothing stores like Benetton, Zara, and H&M. Further south, the street changes name as it bends towards the centre, gathering together the elite brands in the 'Made in Italy' stable: Prada, Armani, and Versace. On weekdays in the afternoon and early evenings and throughout the weekend, the footpaths that run past the shops are almost impassable, filled with people window shopping. At these times, the street edge of this footpath was once the site of a second informal, precarious, and unauthorized geography of commerce. Shadowing the shops that line the footpath, groups of Senegalese men sold counterfeit luxury goods from the small white sheets that they carefully spread on the ground in the periods between the increasingly frequent patrols of police. These men were the visible nodes in a shadow economy that made luxury brands available for everyday middle-class consumption.

Corso Buenos Aires is a site in Milan that was renowned for the regular presence of West African men selling counterfeit luxury goods. With Zona Brera (another well-known informal marketplace), it was targeted under the government's crackdown on the sale of counterfeit fashion, first announced in December 2007. Through attention to Corso Buenos Aires and Zona Brera, this chapter examines how the microdynamics of place mediated the reception of the government crackdown, with its legitimacy varying between different sites across Milan. Drawing on fieldwork conducted in both sites between April 2008 and March 2009, this chapter focuses on the differing relationships between street traders and business owners at each site, and pays attention to how these relationships affected the perception of the crackdown.

FAKE-FAKES AND TRUE-FAKES: COUNTERFEITING 'MADE IN ITALY'

Counterfeit goods sold by street traders in Milan are divided into two main types: 'true-fakes' and 'fake-fakes.' True-fakes are identical in every way to the 'original': they are made with the same materials and same labels. These true-fakes are typically produced as overruns by the factories commissioned with supplying originals to the formal market, and distributed without the rightholder's consent. True-fakes are therefore distinct from fake-fakes, which arrive from China and Eastern Europe. The term fake-fake may refer to inferior copies of an existing model and design. Increasingly, however, it indicates goods that simply have a stolen brand identity. The latter bear no resemblance to officially manufactured goods, in either style or materials, except for the (poorly) counterfeited logos or labels that are attached to them (Hetzer, 2002, p. 304; see also Saviano, 2006).

Whether true-fake or fake-fake, the appeal of counterfeit merchandise depends on the consumer imaginary evoked by the "Made in Italy" brand. Introduced in 1990, the brand certifies the entire production cycle of food and fashion products, guaranteeing that these products have been realized entirely in Italy. According to a press release from the President of the Federation of Italian Fashion, "Made in Italy" evokes an "image of quality and of sophistication," and, furthermore, it is a "status symbol that invokes a particular lifestyle" (Borghi, 2005). The trade in true-fakes and fake-fakes makes the imaginary of 'Made in Italy' (if not the lifestyle) available for mass consumption. Claiming that this accessibility risks undermining the elite connotations of the brand, the Federation of Italian Fashion has actively lobbied government for increased sanctions against the production and importation of counterfeit goods.

THE CRACKDOWN ON THE SALE OF COUNTERFEIT

In December 2007, the Milanese government announced a crackdown on the sale of counterfeit goods by street traders in Piazza Duomo and Zona Brera (Vernuccio, 2006). The crackdown was extended to Corso Buenos Aires in 2008 (ANSA, 2010; Liso, 2009; Vanni, 2008; 2009; 2010a; 2010b). This measure was structurally and politically enabled by the Pact for a Secure Milan, one of 16 pacts that the Italian Minister for the Interior signed with various city governments, between May and July of 2007 (Camera dei deputati, n.d.; Ministero dell'Interno, 2007b; 2007c; 2009). Each of these pacts was based on guidelines that were set out in a 2007 accord, signed by the Italian government and the National Association for Italian Municipalities (ANCI). That accord gave greater flexibility to mayors to introduce legislation and policy to address local security issues, and earmarked funding that could be used discretionally for the implementation of these locally determined policies (Ministero dell'Interno, 2007a; 2008).

The identification of the sale and distribution of counterfeit merchandise as an explicit government target within the pact was not an automatic admission that the general population perceived consumption of counterfeit merchandise to be criminal, or even illegitimate. As Italy's National High Commission for the Fight against Counterfeiting noted in 2005 (Repubblica

Italiano, n.p.), the production and purchase of counterfeit and fake brands were not generally regarded as a "socially and economically damaging crime." The commission therefore argued that one of its "strategic tasks" was to "engender collective knowledge and consensus regarding the harm that is derived from the passive acceptance of this reality and consequently of the necessity to react" (Repubblica Italiano, 2005, n.p.). In other words, it was not enough to criminalize and police the production and distribution of counterfeit goods in law; the commission also had to shape the social perception of these acts.

To underscore how the legitimacy of particular acts may be perceived differently by state and nonstate actors, van Schendal and Abraham (2005) make a distinction between "legality" and "licitness" (p. 18). While legality refers to the "norms and rules of formal political authority," licitness and its counterpart, illicitness, "refer less to the letter of the law than to social perceptions of activities defined as criminal" (p. 18). In the final section of this chapter, I will show how the perception of licitness or illicitness in the sale of counterfeit goods varied according to the microdynamics of place. In doing so, I will illustrate how the Milanese government's crackdown on the sale of counterfeit goods by African street traders came to be seen as more legitimate in some places than in others.

(IL)LEGITIMATE EMPLACEMENTS OF THE 'FIGHT AGAINST COUNTERFEIT'

The position of West African traders in Italy has always been ambivalent. Although the perception of West African traders is generally not overly negative, situational circumstances and differences such as the particular economic structure of a place or the perception of unfair competition can initiate and sustain hostility towards them (Carter, 1997; Riccio, 1999; 2001; 2007). To highlight the importance of place to the changing perceptions of migrants, Bruno Riccio (2001) compared the reception of West African traders in Ravenna and Rimini—two towns on the Adriatic coast, in the region of Emilia Romagna. Both these towns saw the arrival of numerous West and North African traders throughout the 1980s. In Ravenna, a plural economic structure consisting of industry, agriculture, tourism, and trade meant that the trade of Senegalese and Moroccan migrants was not seen as "threatening the wealth of the community" (p. 590). By contrast, in the tourist-oriented trade economy of Rimini, migrant traders were seen as an economic threat, with local business owners organizing frequent demonstrations against them throughout the 1990s. These demonstrations often invoked racist and criminalizing discourses, which were, by and large, absent in the case of Ravenna.

This section builds on Riccio's work, by examining the different responses in Corso Buenos Aires and Zona Brera to the crackdown on the sale of counterfeit goods and migrant street traders. Before the action, both sites shared the daily and visible presence of large numbers of Senegalese street traders, and were therefore explicitly targeted by the Milanese government in their 'fight against counterfeiting." In both sites, the government authorized regular street patrols by local police and financial police, in order to disrupt the patterns of sale. However, the reception of business owners to this crackdown was remarkably different in each site, with restaurant owners in Zona Brera condoning the presence of street sellers, while shopkeepers in Corso Buenos Aires reiterated the government's criminalisation of the traders.

Zona Brera is one of the few remaining historic areas in Milan, and is made up of narrow cobblestone streets that are closed to cars. The area is known for its art galleries and the high-fashion boutiques of Corso Como and Via Montenapoleone, as well as for the small bars and restaurants that began to concentrate in Via Madonnina and Via Fiori Chiari in the 1990s. The tensions created by the transformation of a previously residential district are visible in numerous reports on the area, and are implicated in the distinct reactions of bar owners and residents to the crackdown on the counterfeit trade taking place there.

The opening of bars and restaurants changed the rhythms of inhabitation of this previously residential area, turning the once quiet cobblestone streets into places for late-night public consumption. In parallel with this change, the local residents' organisation began to complain about outsiders (both Italian and non-Italian) coming to the area. These residents complained publically about the misuse of space, citing examples of scooters that would block the narrow streets, or teenagers who would gather in the squares and streets, making noise until late. Others, like Elisabetta Oropallo, the head of the local residents' association, spoke specifically of the arrival of the migrant traders who had turned the area into

> [A] souk where it is impossible to walk, due to the tens of people that undertake illegal activities under everyone's nose. They occupy public land and sell counterfeit goods without the fear of repercussions.
>
> (Battestini, 1998, p. 46)

Over the intervening decade, the residents' association actively lobbied the local government to intervene in the area, progressively building on initial requests that the government exert greater control over licenses and limit the opening times of local businesses, as well as introducing nightly patrols to clear people from the streets (Battestini, 1998). In response, the government initiated intermittent police patrols to target the migrant street traders who, increasingly, came to the area to sell counterfeit merchandise. It was not until 2007 and the signing of the Pact for a Secure Milan that regular, almost nightly, patrols of local and financial police were instigated. To reinforce these patrols, the government also introduced an artisans' market.

This market was explicitly intended to limit the migrant traders' ability to conduct business, by occupying the spaces where these men had previously laid out their sheets and merchandise (Comune di Milano, 2009; Vanni, 2008).

Following the crackdown, the number of traders working in the area diminished significantly, and the patterns of sale of those who chose to remain changed. Owing to the frequent patrols and the increased risk of confiscation, most of the men began to sell with limited or no visible product; their bags, belts, and wallets for sale would be stored somewhere nearby that was relatively safe from confiscation. The way they occupied space—loitering in groups on the edges of the street—acted as a signal to passers-by that they were there to sell. With nothing on display, they would whisper to potential clients, asking if they wanted to purchase a bag or a wallet, before disappearing to their secret stash of merchandise, and selecting potential items for sale. The trade in counterfeit was no longer a casual, leisurely affair; people were unable to browse the merchandise while they strolled through the streets, or as they sat at tables outside the restaurants and bars.

A year after the implementation of the crackdown, local bartenders and restaurant owners declared that their customers had declined by over 50 per cent, with overall takings down by 20 per cent (Vanni, 2008). Although the decline in takings may have also been due to the global financial crisis, the restaurant owners demanded a return of the migrant traders, claiming that their customers had deserted the area following the evacuation of the men who sold bags (Vanni, 2008; 2010a). The comments by local restaurant and bar owners in support of the migrant traders and their return indicate that there was a symbiosis between these formal and informal modes of trade. The Senegalese would set up their sheets according to the opening times of the bars and restaurants, one drawing clients to the other. Furthermore, the hours of the migrant street traders matched those of the bars, and hence only appeared after the fashion boutiques (that often sold the authentic version of their fakes) in the area had closed for the evening. They avoided direct competition with these stores, and minimized any potential antagonisms that might otherwise have emerged. As a result, the local merchants' association came out publically in support of the continuing presence of migrant street traders in Brera:

> We are against any form of illegality, and this is not up for discussion, but amongst my associates, the likability of the mobile traders is undeniable. They are always the same guys. They belong to Brera. They should stop selling counterfeit merchandise, and get themselves a visa and a traders licence.
>
> (Mattia Martinelli, in Vanni, 2010a, n.p.)

As indicated in the above quote, the support of the restaurant and bar owners was complex and multifaceted. In addition to self-interest, local restaurant owners also actively manipulated the Milanese government's discourses of securitisation and criminalisation in order to call for the traders return. One owner claimed that "When they are around the Yugoslav pickpockets stay away," while another said, "Now the streets are more dangerous, and the attendance hasn't improved: it was the ladies of 'respectable' Milan that came to buy fake Gucci bags" (Vanni, 2010a, n.p.).

In each of these comments, an implicit distinction is made between illegal but licit activities (such as trading without a license or the sale of counterfeit goods), and illegal and illicit activities (such as theft and drug dealing), with these activities distributed along racialized lines. Although undocumented, the Senegalese men were seen to attract 'respectable' citizens (and tourists) to the area. Moreover, their skills in vigilantly scouring the crowd for plainclothes police were seen as equally effective against the pickpockets who regularly worked the area. The distinctions made by local restaurant owners were not shared by the local residents; lacking direct contact with the traders, they continued to construe these men as a threat owing to their irregular migration status, and the fact that they did not have a license to sell in the street (Vanni, 2008). In making a clear distinction between acts of theft or violence and the (socially acceptable or generally condoned) act of selling counterfeit goods, the restaurant owners of Brera discursively integrated the Senegalese traders into the economy of the area, even as they continually reiterated that these men were 'illegal migrants.'

By contrast, the merchants from Corso Buenos Aires reiterated the government's own language in order to invite a greater crackdown on unauthorized trade in the street. In October 2009, signs appeared in shop windows along Corso Buenos Aires: "AWAY WITH THE IRREGULARS! ENOUGH DEGRADATION!! YES TO RESPECT FOR THE LAW!!!" The appearance of these protest signs was part of an escalation of ongoing and long-term protests led by the local merchants' associations against unauthorized traders. The association's protests had been ongoing since at least the late 1990s, but they gained momentum (and publicity) in 2007 following the crackdown in Brera, and intensified again—as the sign above indicates—following the semi-permanent occupation of a local piazza by African asylum seekers evicted from their housing (Senesi, 2009; Vanni, 2009). Later protests in Corso Buenos Aires conflated the presence of the asylum seekers with that of the traders (as a result of their shared blackness), while earlier protests were focused on the way in which the traders' occupation of the footpaths impeded window shopping, and disrupted the easy flow of foot traffic up and down the footpath and along the front of the shops (Vernuccio, 2006).

Moreover, the migrant traders were increasingly blamed for petty crime on the street, and were thus included discursively in a continuum of diffuse criminality along with "gypsies, thieves and beggars" (Vernuccio, 2006,

p. 7). The language used by the Milanese government against street traders found widespread resonance in Corso Buenos Aires, with many of the merchants collaborating with the government and police in their fight against counterfeiting and unlicensed trade. This cooperation was made official in 2009, as part of a policy that gave 20 shopkeepers the ability to directly contact the local Carabinieri via mobile phone to alert them of the presence of unauthorized traders, pickpockets, and drug dealers (Vanni, 2009).

Through this policy, the government formalized many of the practices of control and surveillance that were already in operation informally along the street. These informal practices influenced the organisation of unauthorized traders along the street, where decisions on where to sell took into account commuter flows, the amount of traffic received by each store, the relationships that had been established (for better or for worse) over time with local shopkeepers, and the risk of confiscation or arrest. As Djibo (name has been changed) explained:

> D: If I arrive first, I can decide where I want to work [. . .] if I want to work here or there. It depends on the people that are here. [. . .] I can put my stuff down in front of a shop, but this person in the shop has bought lots of stuff from me. When she sees I have beautiful things, she comes and says: "I want this one, this one, and this one, at the end bring me this one. I'm inside, but I'll give you my number." When I leave she gives me the money. When we have come to an agreement, I put the stuff aside. When I finish work I put these in a plastic bag and she gives me the money. But you know something? The people that always break our balls are those from Zara and Oviesse. There is always trouble near Zara and Oviesse. They always call the police.
>
> I: Really? But I always see large groups in front of Zara and Oviesse.
>
> D: There are lots of people here at Zara. Lots of people enter Oviesse and Zara. That's why people put down their stuff there. That is why there is a huge percentage of them there. But that's why, if you aren't sure, you go and put your stuff down somewhere else. Far away from Zara. [. . .] If you are new, you can't be in the front line.

As Djibo indicates, the reactions of local shopkeepers are determined by whether they perceive the traders as economic competition, as a harmless social presence, or as a go-between who can obtain counterfeit goods that can then be sold 'officially' in their stores. The perception of the unauthorized traders as competition is not always related to the sale of counterfeit copies of what is available in official stores. As stated to me by the head of local merchants' association 'Buenos Aires Futura,' the competition posed by traders is understood relative to their ability to offer cheap goods owing to their nonpayment of taxes and rent.

CONCLUSION

Through an analysis of the reactions of Italian merchants and restaurant owners to the sale of counterfeit goods by undocumented migrants in two sites in Milan, this chapter has outlined how place affects the reception of legislation and of government policies to crack down on that sale. In Corso Buenos Aires, the local merchants' association continually lobbied government for greater intervention into unauthorized trade and the sale of counterfeit goods. By contrast, the relationship between unauthorized traders and the local restaurant owners in Brera was reasonably positive, with the latter frequently making the distinction between the irregular but condoned activities of the Senegalese, and the illegal and dangerous activities of other, racialized, migrant groups. This shows that the overlap between illegal and illicit practices of sale is not as clear as the government policy suggests, with the legitimacy of street selling contingent on local contexts.

REFERENCES

ANSA. (2010, June 24). Comprano due cinture taroccate in corso Buenos Aires, multa da 200 euro. *La Repubblica*. Retrieved from http://milano.corriere.it/milano/notizie/cronaca/10_giugno_24/comprano-cinture-taroccate-multa-200-euro-1703261775877.shtml

Battestini, F. (1998, November 9). Brera, il Suk di mezzanotte. *La Repubblica*, p. 46.

Borghi, R. (2005). *Tutelare il Made in Italy dalla contrafazzione*. Retrieved from www.mi.camcom.it/show.jsp?page=614657

Camera dei deputati. (n.d.). *I Patti per la sicurezza*. Retrieved from www.camera.it/cartellecomuni/leg15/RapportoAttivitaCommissioni/testi/01/01_cap25_sch03.htm

Carter, D. (1997). *States of grace: Senegalese in Italy and the new European migration*. Minneapolis: University of Minnesota Press.

Comune di Milano. (2007). *Mercatini a Brera anche d'estate. E gli abusivi stanno a casa*. Retrieved from www.comune.milano.it/portale/wps/portal/searchresultdetail?WCM_GLOBAL_CONTEXT=/wps/wcm/connect/Content Library/giornale/giornale/tutte+le+notizie/rapporti+consiglio+comunale+e+attuazione+del+programma+sicurezza/sicurezza_mercatini+abusivi+brera+lug+07

Comune di Milano. (2009). *Più sicurezza in città: Vigili protagonisti*. Retrieved from www.comune.milano.it/portale/wps/portal/CDM?WCM_GLOBAL_CONTEXT=/wps/wcm/connect/ContentLibrary/giornale/giornale/tutte+le+notizie/rapporti+consiglio+comunale+e+attuazione+del+programma+sicurezza/vicesindaco_sicurezza_2009_anno_consuntivo

Hetzer, W. (2002). Godfathers and pirates: Counterfeiting and organized crime. *European Journal of Crime, Criminal Law and Criminal Justice*, 10(4), 303–320.

Liso, O. (2009, November 1). La tolleranza zero del Comune "multe a chi compra false griffe". *La Republica*, p. 2.

Ministero dell'Interno. (2007a). *Le Misure Legislative per la Sicurezza*. Retrieved from www.interno.it/mininterno/export/sites/default/it/assets/files/15/0619_Misure_pacchetto_sicurezza.pdf

Ministero dell'Interno. (2007b). *Patti per la sicurezza: Firmati a Roma e Milano gli accordi tra governo ed enti locali per contrastare la criminalità.* Retrieved from www.interno.it/mininterno/export/sites/default/it/sezioni/sala_stampa/notizie/sicurezza/2007_05_18_Patti_sicurezza_Roma_Milano_.html

Ministero dell'Interno. (2007c). *Patto per Milano sicura.* Retrieved from www.interno.it/mininterno/export/sites/default/it/assets/files/13/2007_05_18_Patto_per_Milano_sicura.pdf

Ministero dell'Interno. (2008). *Le Misure Legislative per la Sicurezza.* Retrieved from www.interno.it/mininterno/export/sites/default/it/assets/files/15/0619_Misure_pacchetto_sicurezza.pdf

Ministero dell'Interno. (2009). *Norme del Pacchetto Sicurezza e collegati: Sintesi per materia.* Retrieved from www.interno.gov.it/mininterno/export/sites/default/it/assets/files/19/0928_100319_-_Norme_pacchetto_sicurezza.pdf

Repubblica Italiano. (2005). Alto Commissario per la lotta alla contraffazione. Rome: Repubblica Italiano.

Riccio, B. (1999). Senegalese street-sellers, racism and the discourse on "irregular trade" in Rimini. *Modern Italy,* 4(2), 225–239.

Riccio, B. (2001). From "ethnic group" to "transnational community"? Senegalese migrants' ambivalent experiences and multiple trajectories. *Journal of Ethnic & Migration Studies,* 27(4), 583–599.

Riccio, B. (2007). "Toubabe" "Vu Cumpra": Transnazionalita e rappresentazioni nelle migrazioni senegalesi in Italia. Padua: CLEUP.

Saviano, R. (2006). *Gomorrah.* New York: Farrar, Straus & Giroux.

Senesi, A. (2009, October 26). Rivolta in Buenos Aires: "Via abusivi e clochard". *Corriere della Sera,* p. 2.

Van Schendal, W., & Abraham, I. (Eds.). (2005). *Illicit flows and criminal things.* Indianapolis: Indiana University Press.

Vanni, F. (2008, July 21). Brera è diventata zona morta. I commercianti: Ridateci gli abusivi, senza di loro la gente non viene. *La Repubblica,* pp. iii.

Vanni, F. (2009, February 13). Corso Buenos Aires, i negozianti diventano sentinelle contro gli abusivi. *La Repubblica.* Retrieved from http://milano.repubblica.it/dettaglio/corso-buenos-aires-i-negozianti-diventano-sentinelle-contro-gli-abusivi/1589749

Vanni, F. (2010a, April 7). Brera e gli ambulanti abusivi "Tuteliamoli, portano clienti". *La Repubblica.* Retrieved from http://milano.repubblica.it/cronaca/2010/04/07/news/brera_e_gli_ambulanti_abusivi_tuteliamoli_portano_clienti-3162735/

Vanni, F. (2010b, June 17). False griffe, vigili in borghese per i clienti. I negozianti: "Grazie per aver ripulito la via". *La Repubblica.* Retrieved from http://milano.repubblica.it/dettaglio/false-griffe-vigili-in-borghese-per-i-clienti-i-negozianti:-grazie-per-aver-ripulito-la-via/1788918

Vernuccio, M. (2006, October 25). Noi commercianti, costretti a pagare i vigilantes contro gli abusivi. *Corriere della Sera,* p. 7.

17 The Importance and Necessity of the Informal Market as Public Place in Delhi

Ranjana Mital

Long before the commercial centre or the shopping mall, there were bazaars that stretched on both sides of the main street of the city. The still bustling Chandni Chowk market of 17th-century Shahjahanabad, Delhi, tells this story. For centuries, the pattern has been similar in towns and cities throughout India. Folklore, popular songs, and romances often refer to the great excitement surrounding the arrival of the bangle-seller, or an outing to the annual fair. Rabindranath Tagore has immortalized the 'Kabuliwala' in the story by the same name. Adapted for at least three feature films, this well-loved tale is about an Afghan who came from Kabul to Calcutta every year to go door to door, selling the dry fruit he brought with him from his homeland. Shops lined important streets and provided the impetus for vibrant social and commercial activity. These shops were an intended and essential part of the formal city. This built fabric, while giving the city's architecture a character, also nurtured and sustained an important social space. Indeed, it possibly illustrates a strangely egalitarian or democratic urban-space sharing system in the sense that the governance in most of these instances was autocratic.

This chapter looks at the weekly informal urban street markets in Delhi, which have, along with travelling salesmen, kept the common people's homes supplied. It interrogates the importance of these 'social events' which enable a judicious, shared use of urban space and of public resources to provide democratic life chances. It also draws attention to the fact that these informal urban street markets and their vendors provide for the city far more than the city provides for them (Mital, 2005). The overarching objective of this chapter is to show that the informal urban street markets are imperative for the social well-being of Delhi city. This will be done by briefly describing the typical markets of Delhi in terms of their merchandise, clientele, organization, vendor profiles, and problems, and acknowledging the potential of the National Policy for Street Vendors in directing city and street market development. The chapter identifies the roles that architects, planners, city administrators, and city dwellers play, and directs attention to the potential these markets have to create sustainable, sociable, and multi-functional places. The chapter concludes by underlining the importance of the informal markets as an integral part of the socio-physical fabric of a city, an importance that extends its financial role.

THE WEEKLY MARKET

When it comes to the 'not-so-elegant' informal markets, public opinion and action varies from concerned awareness programmes initiated by nongovernmental organizations (NGO) to demands for these markets to be cleared out in the cause of city security and beautification. In urban India informal markets comprise a range of retailing, hence social options: from the lone travelling vendor moving along the street selling anything from trinkets to fresh vegetables and fruit to the 'organized' informal or weekly markets that consist of hundreds of stalls. Hashmi (2007) suggests that these weekly markets owe their beginnings to the inevitable village in the environs: wherever you find a weekly market he says you will find an old Delhi village. The villages are long gone, having been overrun by real-estate development. What remain are mostly haphazard, unsafe, and overpopulated areas that flout all building norms. These markets have persisted in the face of such development, and even flourished, which only underscores their necessity to the buyer as much as the seller for both economic and social reasons.

The informal bazaars or markets are of three distinct kinds; the weekly bazaar, the daily bazaar, and travelling salespeople. This chapter focuses on the weekly bazaars. They offer not just the daily grocery essentials but also items such as ready-to-wear garments or large suitcases and trunks. Some people come to the market only to enjoy the sweets and savouries being sold in the street-food section of the market. A visit to the weekly market is often a social outing with family or friends. Shopping at these bazaars is also about bargaining. Rarely would the vendor's price be taken as fixed or final. The customer invariably challenges the vendor in mock outrage or downright aggression and depending upon the skill of the vendor either wins or loses. Often the actual transaction is seemingly the least important of the little drama that is enacted. Customers, strangers to each other, thus bond for a few moments with each other and with the main protagonist, the vendor, then go their separate ways. However, not before creating a social event and place on the dusty sidewalks of a bustling city. It is not only the goods but the vendors and this sociability that are magnetic. A successful vendor knows instinctively that he has to draw and hold customers, and is usually quite skilled in the art of vendor–customer sociability.

A similar spectacle plays out with the lone vendor on his cycle. He cycles slowly around a residential area and sooner or later is hailed by a pedestrian. He stops and quickly has an animated little group around him. Some queries, some small talk amongst neighbours, sometimes a transaction, and the vendor is on his way once more—the social event that 'coloured' the plain street already a memory. Travelling vendors and informal markets generate an important social activity. With the increasing 'space crunch,' escalating land values, and resulting spatial inequities these social events serve as valuable people-place providers. Indeed, this may be the informal markets' most important contribution to the social well-being of any city.

Walking down any one of these markets, one is struck by the diverse clientele these places boast. The majority of customers are from the informal sector—casual workers, labourers, domestic help, and other nonskilled, poorer sections of society, with little or no education. The make-up of these markets ensures very competitive pricing. Locally produced vegetables are invariably fresher and cheaper than in the regular stores. There are also people who are better off but who patronize these markets because they are convenient. College students and young professionals scour some of these bazaars known for selling export-surplus clothing at throwaway prices, as do well-to-do matrons out on the lookout for a bargain buy.

The National Policy on Urban Street Vendors, first published in 2004 and revised in 2009, defines "street vendor" as a "person who offers goods or services for sale to the public in a street without having a permanent built-up structure." It establishes three categories: stationary, peripatetic, and mobile (Government of India, 2009, p. 3). The people who run the stalls at these informal urban street markets are frequently poor, illiterate, and 'unskilled.' Rural poverty in India forces villagers to migrate to the city. The Census 2001 figures show that the net migration to Delhi was 1.7 million people in the decade preceding 2001. Of this, the majority was male, from neighbouring states, and in search of work or employment (Census of India, 2001, p. 24). Once in the city, jobless migrants, usually uneducated but with a will to survive, seek absorption into the informal sector. With some resources drawn from pooled savings, maybe a loan from a local moneylender, and a load of grit, they turn to vending. My research in some of Delhi's weekly markets shows that for many of the vendors this work has become a way of life, with some having been at it for as long as 30 years. Apart from migrants, there are instances of people losing their jobs in the formal sector and finding their livelihood here.

While the vendors can be described as members of the unorganized or informal sector by city authorities, economists, and planners, the manner in which the whole phenomenon of the informal market functions is organized—formal and informal arrangements are intertwined. Political theorist Partha Chatterjee (2008) explains how the informal sector often relies on large, effective, and powerful management systems usually headed by local politicians. According to Chatterjee, using political leverage to garner governmental support, the informal sector has been able to develop a "fairly stable and effective" noncorporate economy that sustains the livelihood of the urban poor (n.p.).

At the informal markets, the leader is known as the *pradhan* or chief. He is the link between the vendors and the municipal authorities and the local police. He arranges to pay the *teh bazari* which is a "tax collected on a daily basis by local authority from small traders for selling their items in a weekly market or any other public place," as well as the *hafta*, a "payment, mostly illegal, made on a weekly basis to officials in authority by petty industrialists, traders, or slum dwellers" (Kundu, 1999, n.p.). The *pradhan* also plays the role of arbitrator in the event of disagreements between vendors, and takes on the role of protector against a bully policeman, a municipal corporation official, or even a particularly nasty customer. Wielding such authority, the *pradhan* invariably exploits his position to get the best for himself and his close associates. Harassment by the *pradhan* is common, although possibly less exploitative than the systems uncovered by Sharma (2000) in Mumbai. On a bad day, a vendor could be at the receiving end of some bullying or injustice at the hands of the *pradhan* and not have much to keep after paying off their daily rental. Bad luck could get worse if a municipal functionary along with the local police decided to 'inspect' and confiscate 'illegal' handcarts and all the merchandise. Illegal here implies those who do not have a *teh bazari* licence and who may or may not have paid the *hafta*. A vendor's life is insecure by any standards. Yet vending is the most lucrative of activities open to the urban poor, and therefore vendors bear the harassment and humiliation.

THE SUNDAY MARKET AT RAMA KRISHNA PURAM, NEW DELHI

The Sunday Market at Rama Krishna Puram (RKP) is a typical example of a weekly informal street market. It stretches along a link between two busy roads. As Roever (2012) points out, it is the location that is most important. To be profitable an informal urban street market has to be en route: strategically located in urban areas with steady pedestrian flows. In the case of the weekly market in Delhi it is usually located just off main roads or between two intersections, in places that assure the vendors of a large catchment of passersby.

The Sunday Market is made up of approximately 200 stalls selling apparel, shoes, bags, blankets, plastic ware like buckets, bowls, toys, watches, rugs and carpets, cushion covers and bed linen, coats and woollens, steel utensils and iron griddles and woks, as well as *bindis*, bangles, and other cosmetics. Vegetables, fruit, dry pulses, and spices are a constant, as are the stalls selling street-food that may be washed down with hot tea or cold mint-flavoured fresh-lime water. The Sunday Market moves on a daily basis. It is the 'Monday Market' at Saket, the 'Tuesday Market' at Bhogal, and the 'Wednesday Market' at Kalkaji. 'Thursday Market' and 'Friday Market' are set up at Masjid Moth and Govindpuri, respectively, while the 'Saturday Market' occurs at Lajpat Nagar.

Each of these markets is unique, with the vendors knowing what will be popular on a particular day or in a particular locality, and varying their stock accordingly. The street knowledge and skill of gauging consumer demands, and then supplying them at the least possible cost has allowed the weekly informal bazaar to prevail through political changes and economic upheavals over centuries. The informal market's biggest advantage is that it sells almost anything at the cheapest possible rates and yet returns a profit to the vendor. While profit margins must be kept low enough to meet customer satisfaction, the vendor can derive comfort from the fact that their expenditure on infrastructure is negligible and they hardly ever suffer any losses.

The markets usually set up on rented tables provided with electric lighting. Location of tables is decided by consensus among the vendors, or by the *pradhan*. The workers from the agency hired to set up the stalls have mastered the art of laying out the tables and goods and just as quickly packing them up again. Hectic activity starts with the setting up of the market in the early afternoon, peaks in the evening after office hours when the maximum number of customers arrive, and then gradually winds down to pack-up time, at approximately 10:00 p.m., weather permitting. Earlier, the vendor will have spent the best part of the morning bargaining with wholesalers for their merchandize. Goods are sourced from the various wholesale markets in the city and carted to the location for the day by shared transport. Cooperation is the key to keeping overheads down. Thus, across the market vendors, especially those selling similar goods and hence going to the same wholesale market, will for example pool resources to hire transport or rope in family members to assist with the stalls. With the markets being on the move there is little storing of goods in large quantities. Only as much as will easily sell or can be stored for a few days, with the rest of their belongings, is procured. In any case, vendors do not have the capital to invest in large amounts of stock at any given time.

OCCUPATIONAL HAZARDS ON THE STREETS

It is difficult to stop this city phenomenon. It is not just the persistence of the vendors but also the throngs of customers at the weekly markets who would convince any skeptic that the continuation of these informal urban street markets is important. However, the temporary and mobile characteristics of the informal urban street market make it the first target of government-driven demolition drives. For example, the Monday Market site at Saket was slashed to almost half its size to make way for a complex of buildings housing the district courts, leaving many people in the lurch.

While the laws have always allowed for some sort of informal vending after procuring the requisite licences, the police-municipality nexus most often prefers to keep it to casual, oral permission, without formal receipts but after some monetary transaction. The reasons are many, the most important being that the number of vendors wanting licenses far exceeds the allowable number. This arrangement is accepted because it allows for the newcomer to set up without a license and to delay paying *hafta* until they have established themselves. However, it also means that when the authorities decide to flex muscle, the vendors have no *locus standi* against the ensuing harassment. Threats and instances of cruelty where carts are taken away or goods confiscated by the police are commonplace, as seen from reports from NOIDA (New Okhla Industrial Development Authority) (Sekar, 2008), but for most vendors, there is not an alternative.

The greatest threat comes from the city developing authorities themselves when they embark upon 'city-beautification' drives. Given that temporary and mobile informal urban street markets are perceived as shabby, down-market, and dispensable by the city authorities, they are among the first to be cleared to make way for a Delhi that is a 'World Class City.' Jhabvala (2000) calls for a change to this perception: "beauty need not mean the exclusion of large sections of the population" and gives us the choice of perceiving the vendor as part of our culture to be 'preserved and upgraded' or like a 'pot-hole in the road to be removed as soon as possible' " (n.p.).

THE NATIONAL POLICY FOR STREET VENDORS

The Model Street Vendors Bill when passed by Parliament will provide for social security and livelihood rights to the street vendor. It addresses the issues considered by the earlier National Policy (2009) that provides visibility and a voice for the vendors. The National Policy for Street Vendors is unique in that it is one of the few policies in the world for street vendors. While it is not yet law, there have been some landmark judgments given by the Supreme Court that serve to uphold the fundamental rights of street vendors.[1] Sinha and Roever (2011) believe the policy will serve as a model internationally as it "prioritizes inclusive urban planning processes, with a focus on giving voice to street vending associations" (p. 5).

The overarching objective of the policy is "to provide for and promote a supportive environment for the vast mass of urban street vendors to carry out their vocation while ensuring that their vending activities do not lead to overcrowding and unsanitary conditions in public spaces and streets" (Government of India, 2009, p. 4). The specific objectives are as follows:

a. To give vendors a legal status;
b. To provide civic facilities for appropriate use of identified spaces;
c. To provide transparent regulation with respect to allotments of licences and space;
d. To promote, where necessary, organizations of street vendors to facilitate their collective empowerment;
e. To set up a participatory process that involves the local authority, planning authority, police, associations of street vendors and resident welfare associations, NGOs, representatives of professional groups, trade and commerce, banks, and eminent citizens;

f. To promote norms of civic discipline by institutionalizing mechanisms of self-management and self-regulation;

g. To promote the access of street vendors to services such as credit, skill development, housing, social security, and capacity building (Government of India, 2009, pp. 4–5).

The policy is soon to become a law protecting the rights of the vendors, and the Master Plan for Delhi 2021 makes it mandatory to include designated areas for street vending in the layout plans for the 33 wards taken up for the pilot exercise in the city. Work has already begun on the layout plans with many institutions in the city having been entrusted with preparing the details for the various wards. These measures will no doubt bring greater attention to street vending in the future.

Cooperation works better than working at cross-purposes, a truth illustrated in Bhubaneswar, the capital city of Orissa state (Kumar, 2012). Partnering with member-based organizations, city planners were able to develop rules that the vendors could be relied on to abide by, thereby creating a self-regulating community which circumvented the need for police vigilance. Roever (2012) attributes the success of the Bhubaneswar project largely to the fact that they recognized the natural markets of the city and located the informal markets there.

Two decades separate the formation of the National Association of Street Vendors of India, in September 1998, and the formulation of the National Policy for Street Vendors, in 2009. Giving the informal market its fitting recognition was long overdue. Although the move by city planners to provide dedicated areas for street vending is laudable, providing for physical infrastructure might, paradoxically, erase the very identity of the informal market. There must be a distinction drawn between providing for a permanent shopping street and allowing for informal weekly markets as street happenings. They transform the street with their activities for a few hours and then leave the street to its function of traffic circulation, without as much as a trace of its existence. A little exaggeration here may be excused in the interest of the larger argument. Like most large public gatherings, street markets leave in their wake a lot of litter for the municipal cleaners to tend. Returning to the argument against building for informal markets that some cities are engaging in, this chapter argued that by providing markets with so much as a low, open-to-sky platform, the informality and impermanence is lost. It would not long remain a weekly bazaar but would be transformed into a regular everyday one, gradually to become permanent and eventually driving newer street vendors away.

Planners and architects would do much better by providing space but not building on it, letting the informal market create its own physical setting, as well as establish a social one. This is the unique value of the informal market—its capacity to create a popular place for people out of virtually nothing and, importantly, leaving an area for another kind of socio-physical intervention. Poignantly, like those manning the stalls or carts, these markets give to the city more than they get from it. This must be acknowledged and encouraged.

THE IMPORTANCE OF THE INFORMAL URBAN STREET MARKET

Until the National Policy for Street Vendors was formulated, those who recognized the importance of the informal urban street markets could only stand on the sidelines and wring their hands in despair as autocratic administrators swept off the face of the city all that was perceived as ungainly and a 'blot' on its 'world class' status. With the legal rights to come, and the number of advocates for the informal markets steadily increasing, the informal urban street market will be held in greater esteem in the formal economy. The policy recognizes that informal urban street markets are integral to the city and that street vending as a means to a livelihood is a fundamental right. In addition to recognizing street vendors as an "integral and legitimate part of urban retail trade and distribution systems for daily necessities of the general public," the policy also acknowledges the role of the informal market in: providing employment opportunities to an unskilled majority and thus "combating unemployment and poverty"; sustaining small-scale production units and/or ancillary service providers to the market; and "providing valuable services to the urban masses" (Government of India, 2009, p. 2).

The National Policy for Street Vendors is a landmark document. However, it still looks at the informal sector, street vendors in particular, in a supercilious way. Despite its recognition of the positive role played by informal urban street markets, the policy 'talks down' to the urban poor. It defines and celebrates informal urban street markets' role as serving the 'urban masses' and helping to alleviate poverty, yet fails to acknowledge the enormous role it plays in creating a platform for that 'elusive' sustainable integrative inclusive social system (Aggarwal, 2011). At the weekly market concepts like 'social integration' and 'inclusive' find meaning and can be experienced, however short-lived. Yet the market belongs predominantly to a particular social group. Vendor and customer, together, create out of streets and interstitial spaces, successful places for people usually with very little access to public places, thereby increasing the much-needed public domain without building any edifice or encroaching on anybody's land. For this reason too, the informal market is essential to the social well-being of the city.

It might be argued that at this juncture, when the policy is just beginning to be implemented and the Government of India is being lauded for supporting such a policy, any kind of criticism is unhelpful. The contention here, however, is not about the importance of the National Policy for Street Vendors or about its tremendous potential in securing for the vendor his or her security and dignity. It is about failing to look at the informal market as heritage, as place-maker, and as a self-generating sustainable social event that goes beyond the economic.

The importance of the informal market and street vendors as a continuing tradition notwithstanding, the increasing social and spatial inequities in our cities today compel us to look at the informal market not just as an opportunity for employment for the poor and jobless, but as a highly successful place for people. The high degree of social activity generated by the market is a manifestation of the success of the 'place' even as it overrides considerable physical shortcomings of the site as a place for people.

The wide mix of people in terms of economic class who frequent these markets makes the informal bazaar a natural meeting ground for the various economic groups. However momentary, the exchanges (sometimes wordless, as the smiles between two women watching a child's antics or animated as with two shoppers comparing notes about prices or commodities) are recognized as possibly the most valuable fallout of an informal market event in the neighbourhood. Temporarily becoming part of the informal and ephemeral market, one begins to recognize the other, generating mutual respect. While the National Policy on Street Vendors hails street vending and informal markets as providing an essential service to the city, it fails to acknowledge that this service is much more than selling consumer goods at extremely reasonable rates. The weekly, informal market is the very space for creating an integrated, egalitarian, urban social fabric. Here is where divisions between the two halves of the city are blurred. Here is where public space is truly public. This is its biggest contribution to the city.

NOTE

1. A case in point is the 1989 case of Ms Sodan Singh vs New Delhi Municipal Council where the Supreme Court of India ruled that "small traders on sidewalks can considerably add to the comfort and convenience of the general public [. . .] the right to carry on trade or business mentioned in Article 19 (1) of the Constitution on street pavements, if properly regulated, cannot be denied on the ground that the streets are meant exclusively for passing or re-passing and no other use."

REFERENCES

Aggarwal, B. (2011). *The elusive inclusive city: Temporary informal markets in the modern Indian city taking the case of Delhi.* Unpublished dissertation, School of Planning and Architecture, New Delhi.

Census of India (2001). *Data highlights: Migration tables.* Retrieved from http://censusindia.gov.in/Data_Products/Data_Highlights/Data_Highlights_link/data_highlights_D1D2D3.pdf

Chatterjee, P. (2008, June 13). *Democracy and economic transformation* [Web log comment]. Retrieved from http://kafila.org/2008/06/13/democracy-and-economic-transformation-partha-chatterjee/

Government of India, Ministry of Housing and Urban Poverty Alleviation (2009). *National policy on urban street vendors.* New Delhi: Government of India. Retrieved from http://mhupa.gov.in/policies/StreetPolicy09.pdf

Hashmi, S. (2007, August 6). *The hafta bazaars of Delhi* [Web log comment]. Retrieved from http://kafila.org/2007/08/06/the-hafta-bazaars-of-delhi/

Jhabvala, R. (2000). Roles and perceptions. *Seminar, 491.* Retrieved from www. india-seminar.com/2000/491/491%20r.%20jhabvala.htm

Kumar, R. (2012). *The regularization of street vending in Bhubaneshwar, India: A policy model.* WIEGO Policy Brief (Urban Policies) No. 7. Women in Informal Employment Globalizing and Organizing. Retrieved from http://wiego.org/sites/ wiego.org/files/publications/files/Kumar_WIEGO_PB7.pdf

Kundu, A., & Basu, S. (1999). Words and concepts in urban development and planning in India: An analysis in the context of regional variation and changing policy perspectives. Retrieved from www.unesco.org/most/p2basu.htm

Mital, R. (2005). Not homeless but houseless in Delhi. *Design Philosophy Papers, 3,* 12–18.

Roever, S. (2012, November 13). Governance: How street vendors and urban planners can work together. *The Global Urbanist.* Retrieved from http://global-urbanist.com/2012/11/13/vendors-planners-work-together

Sekar, H.R. (2008). *Vulnerabilities and insecurities of informal sector workers: A study of street vendors.* New Okhla Industrial Development Authority: V.V. Giri National Labour Institute.

Sharma, R.N. (2000). The politics of urban space. *Seminar, 491.* Retrieved from www.india-seminar.com/2000/491/491%20r.n.%20sharma.htm

Sinha, S., & Roever, S. (2011). *India's national policy on urban street vendors.* WIEGO Policy Brief (Urban Policies) No. 2. Women in Informal Employment Globalizing and Organizing. Retrieved from http://wiego.org/publications/ india%E2%80%99s-national-policy-urban-street-vendors

Contributors

Yusuf Abdulazeez is a Lecturer in the Department of Sociology, Usmanu Danfodiyo University, Sokoto, Nigeria, where he teaches on social policy and welfare studies, social statistics, and African studies.

Suresh Bhagavatula is an Assistant Professor at the N.S. Raghavan Centre for Entrepreneurial Learning (NSRCEL), Indian Institute of Management, Bangalore.

Micol Brazzabeni is Associate Research Fellow (Foundation for Science and Technology) at CRIA-IUL. Her current research with Portuguese *cigano* families addresses processes of commodification in contexts of formal/informal economies in urban open-air markets, and relationships between spaces, emotions, and social and institutional suffering. Her fieldwork with an indigenous population of Minas Gerais (Brazil) focused on the process of creating and developing the indigenous school and the training course for indigenous teachers.

Asli Duru is a Researcher with the Emmy Noether Research Group at Ludwig-Maximilians-Universität München, working on the project "From the Oriental to the 'Cool' City. Changing Imaginations of Istanbul, Cultural Production and the Production of Urban Space." Her research interests include critical feminism, relational political economy, urban studies, cultures of food, and environmental movements.

Clifton Evers is currently a Lecturer in Cultural and Media Studies at the University of Nottingham Ningbo China. He has conducted research for government departments, elite sporting organizations, community groups, media outlets, and private industry. Clifton's work has been published as publicly available government reports and as academic articles in international peer-reviewed journals such as *Social & Cultural Geography*, *International Journal of Communication*, *Sport and Social Issues*, and *Cultural Studies Review*.

Lelia Green is Professor of Communications at Edith Cowan University in Perth, Western Australia. The ethnography in this book was undertaken as part of an ECU-supported sabbatical in 2013. Her most recent book is *The Internet* (Berg, 2010). She is Chief Investigator on the AU Kids Online project, funded by the Australian Research Council Centre of Excellence for Creative Industries and Innovation.

Kate Hepworth is a cultural geographer who works on migration, citizenship, and labour. Her 2012 PhD dissertation for the University of Technology, Sydney, was titled *Encounters with the Clandestino/a and the Nomad in Milan: Irregularisation, Securitisation and 'Illegitimate Outsiders' Through the 2008 Italian Security Package.* Kate is currently a visiting fellow at the University of Technology, Sydney, Australia.

Nashaat H. Hussein is an Assistant Professor of Childhood Sociology at Misr International University, Cairo, Egypt. He has published research in the *Journal of the History of Childhood and Youth*, the *International Journal of Sociology and Social Policy*, and recently published in *Families, Relationship, and Societies* on motherhood in Egypt.

Kiran Keswani is an architect based in Bangalore. She is currently pursuing her doctoral studies in Urban Design at CEPT University, Ahmedabad.

Maša Mikola teaches anthropology, sociology, and international studies at the University of Melbourne and RMIT University, Melbourne. She is also a researcher on a project looking at the experiences of temporary migrants in labour markets in Australia. She has a doctorate in Intercultural Studies. Her interests are in the fields of urban anthropology, migration, nationalism, and citizenship.

Ranjana Mital is a Professor at the Department of Architecture, School of Planning and Architecture, New Delhi, and teaches design and history of architecture. She has a PhD in Architecture (from the School of Planning and Architecture, New Delhi), and a B.Arch. (from the University of Delhi). Her concern for the worsening quality of life in cities in India informs much of her research. She has published in national and international peer-reviewed journals.

Tony Mitchell is a Senior Lecturer in Cultural Studies at the University of Technology, Sydney. He is the co-editor of books including *Global Noise: Rap and Hip Hop Outside the USA* (2001), *Home, Land and Sea: Situating Music in Aotearoa New Zealand* (2011), *Sounds of Then, Sounds of Now: Popular Music in Australia* (2008), and the author of *Dario Fo: People's Court Jester* (1999), as well as numerous articles on popular music and film around the world.

Sundramoorthy Pathmanathan is an Associate Professor of Criminology and faculty member in the School of Social Sciences at Universiti Sains

Malaysia, Penang, where he teaches criminology and research methods/ methodology. Currently, he is the Deputy Dean of Students Affairs and Networking. He has many published articles, workshops, and conference presentations.

Emily Potter is a Senior Lecturer in the School of Communication and Creative Arts at Deakin University, Melbourne. She researches in the areas of literary and cultural studies, with a particular focus on questions of environmental practices and politics. She is the co-author of *Plastic Water* (with Gay Hawkins and Kane Race), and co-editor of *Ethical Consumption: A Critical Introduction* (with Tania Lewis).

Kirsten Seale is Adjunct Fellow at the Institute for Culture and Society at the University of Western Sydney. Kirsten's current research investigates the role of street markets in the (informal) production and communication of place in urban spatial economies. She is the author of the forthcoming *Markets, Places, Cities* (Routledge, 2015). Recent work has appeared in *Media International Australia*, *Cultural Studies Review*, *Text*, *Streetnotes*, and *Meanjin*.

Daisy Tam is a Research Assistant Professor in the Department of Humanities and Creative Writing at Hong Kong Baptist University. Daisy received her PhD in Cultural Studies from Goldsmiths, University of London. She is an ethnographer whose research interests include ethical food practices, urban farming, migrant workers, communities, and cultural theory. Publications include "Towards a Parasitic Ethics" (forthcoming *Theory, Culture and Society*); and 在上海街頭跳舞 "Dancing in the Streets of Shanghai" (2011) and 家常便飯 "On Collective Memory and Identity" (2010) in 本土論述 *Journal of Local Discourse*.

Khalilah Zakariya has an academic background in landscape architecture and tourism planning. Her research explores these interrelated areas, with an interest in how we design and experience local places in the city. She obtained a Bachelor of Landscape Architecture from International Islamic University Malaysia, a Master of Science in Tourism Planning from Universiti Teknologi Malaysia, and a PhD in Architecture and Design from RMIT University, Australia. Her current research projects include mapping informal activities at urban squares and examining marketplaces in Malaysia.

Dunfu Zhang is a Professor in the School of Sociology and Political Science at Shanghai University. He was trained in the Institute of Sociology and Anthropology, Peking University. His areas of interests are the sociology of consumption, consumer culture, and urban China.

Index

For Product Safety Concerns and Information please contact our EU
representative GPSR@taylorandfrancis.com
Taylor & Francis Verlag GmbH, Kaufingerstraße 24, 80331 München, Germany

www.ingramcontent.com/pod-product-compliance
Ingram Content Group UK Ltd.
Pitfield, Milton Keynes, MK11 3LW, UK
UKHW020941180425
457613UK00019B/502